Imaging the Cheops Pyramid

SOLID MECHANICS AND ITS APPLICATIONS
Volume 182

Aims and Scope of the Series

The fundamental questions arising in mechanics are: Why?, How?, and How much? The aim of this series is to provide lucid accounts written by authoritative researchers giving vision and insight in answering these questions on the subject of mechanics as it relates to solids.

The scope of the series covers the entire spectrum of solid mechanics. Thus it includes the foundation of mechanics; variational formulations; computational mechanics; statics, kinematics and dynamics of rigid and elastic bodies: vibrations of solids and structures; dynamical systems and chaos; the theories of elasticity, plasticity and viscoelasticity; composite materials; rods, beams, shells and membranes; structural control and stability; soils, rocks and geomechanics; fracture; tribology; experimental mechanics; biomechanics and machine design.

The median level of presentation is the first year graduate student. Some texts are monographs defining the current state of the field; others are accessible to final year undergraduates; but essentially the emphasis is on readability and clarity.

For further volumes:
http://www.springer.com/series/6557

H.D. Bui

Imaging the Cheops Pyramid

Springer

H.D. Bui
Ecole Polytechnique, LMS
Palaiseau
France
hdbui37@yahoo.fr

ISSN 0925-0042
ISBN 978-94-017-7868-8 ISBN 978-94-007-2657-4 (eBook)
DOI 10.1007/978-94-007-2657-4
Springer Dordrecht Heidelberg London New York

Printed on acid-free paper

Springer is part of Springer Science+Business Media (www.springer.com)

<u>Photo</u>: The Cheops pyramid and visitors Joanna, Oriane and Theo (December 2010)

Preface

This book is the result of an initiative of H.D. Bui. This initiative has two merits:

- On the one hand, it gives us an opportunity to recall the skills of engineers and researchers from EDF mobilized around innovative projects in a programme called the *Technological and Scientific Sponsorship,*
- On the other, it allows us to pay tribute to two departed engineers: *Pierre Deletie* Geologist of the Equipment Division of EDF, Geology Department, whom we called "*homme de terrain*" ["*man of the earth"*], and *Jacques Lakshmanan,* Director of CPGF (Compagnie de Prospection Geophysique Francaise), who defended his PhD thesis on the Cheops Pyramid at the University of Nancy.

We have gathered in this posthumous tribute two men who were outstanding representatives of the links that, twenty five years ago, united experts from the "Big House" and talented engineers of a small « Company ».

As to the aims of this work, it is simple and clear to the author; we hope it will be also for readers. The first is to show that beyond the many theories more or less esoteric on the Great Pyramid, only rigorous scientific analysis based on the best technology can lead us gradually to an understanding of its design and construction. More generally, the process undertaken to understand its unusual construction shows that archaeologists need help from other scientific communities (geophysicists, computer scientists, mathematicians, chemists, *etc*). Scientific scholars (archaeologists, historians, *etc*) still cannot benefit because they are too few and scattered. Progress comes only as we agree to enact and implement sponsorship, combining the skills of their engineers and the power of technologies developed.

It was this "synergy", "working together to a common goal" that led to a new form of Technological and Scientific Sponsorship in 1985. Three projects conducted on the Pyramid of Cheops were the first steps of this great scientific adventure!

The Birth

The birth of this form of action was the result of chance. It all started in 1983 with the restoring of a ferrous alloy canon taken out of the wreckage of the *Slavia Rossii*, a ship belonging to Catherine II which sunk in 1790 near the Island of Levant in the Mediterranean. Researchers from our St Denis electrochemistry group had ways to restore metal objects emerging from underwater excavations, by removing destructive chloride.

The first tangible result was wide media coverage that brought us many requests for help. Unfortunately, many of these requests were for financial support; would be researchers rated financial help higher than access to technological innovation. Also, when Prince Napoleon asked us our financial support for his diving project on the wreck of the *Orient*, the Admiral flagship of Napoleon's fleet sunk by Nelson off Aboukir, we replied that there were better things to do than giving money: to restore objects that emerged from the wreckage.

It was the response to these shipwreck restorations that led to the birth of Technological and Scientific Sponsorship.

Two centuries after Napoleon Bonaparte's expedition, we left the wreckage of the *Slavia Rossii* off Toulon and turned to Egypt. Here, objects removed from the wreck were treated at the Museum of Alexandria, where we installed a small laboratory. Some Egyptians were interested in our processes; we gave an introductory seminar in which we defined this new form of sponsorship, replacing financial support by the transfer of technology.

Technological & Scientific Sponsorship and the Cheops Pyramid

This early work in Egypt did not go unnoticed. In April 1986, we received a request from Mr Guillemin, Division of Cultural, Scientific and Technical Affairs of the Department of Foreign Affairs of France. We were asked to confirm or invalidate the assumptions made by two French architects, MM Gilles Dormion and Jean Patrice Goidin on the existence of unknown passages (entry, corridors,..) leading to true burial chambers, the current visible passages being only lures to deceive potential thieves. Their arguments, contained in the book entitled *« Kheops, nouvelle enquête, » (Editions recherche sur les civilisations, 1986)*, were based on observations of a number of anomalies or, if you prefer, some features of the construction, especially concerning certain oversizing feature in some stones of the wall of the King's Chamber, corridors to the Queen's Chambers, the floors *etc*.

In short, a modern architectural vision coming to the aid of a silent story without signs and scripts to solve the mystery of the pharaoh Cheops and, more importantly, the idea of a different yet unexplored path leading to an inviolate tomb suggested we were on the verge of a great discovery! But this was not our motivation. We knew that in fact the builders of the Great Pyramid were remarkable architects. This was enough to convince us that the preparation of the foundations of this gigantic work, outside and inside, was made by the same techniques used today to build our great works: horizontal soil stripping, filling voids with materials.

But we also knew that these architects were *priests* who did not obey a strict architectural logic and even less design rules that define current modelling of structures. Despite our hesitation, we agreed to try a non-destructive method of verification of the assumptions. We did this in the name of scientific ethics which imposes that no assumption should be rejected a priori. But we have limited our expertise to two points of the Dormion and Goidin thesis. First, an essential element, the alleged existence of a strange room *to the North* and *at the level of the current King's Chamber* and then a problem considered secondary, the presence of an anomaly in the construction of the corridor leading to the Queen's Chamber.

We had to consider what was the best way to solve the problem. We first explored the idea of a radar, but the expert's report showed us that no transducer system (emitter-receiver) would be able to locate a cavity through too large heterogeneous stones and, more importantly, in a relatively wet medium. We also considered using the methods of seismic tomography; but the risks of landslides were too large and the alternation of squared and faced stones and roughly hewn quarry stones used for the filling make it difficult to interpret acoustic signals recordings. Also, as since we were not asked to specify the exact form and the size of the cavity, but only its existence, we used microgravity measurements.

The exact content of this technology, its interest, its scope and limitations are specified in the chapters that follow. Measuring tiny variations in gravity, by an ultra-sensitive scale, can reveal the existence of density variations and therefore unknown cavities. Geologists from the Division of Equipment of EDF commonly used these methods to study the basement of future settlements (dams, power plants, *etc*). And they were not allowed to make any mistakes. In addition, our usual partner CPGF had particularly efficient apparatus.

All the members of these two teams had a proven expertise in both the measurements and the data processing. The implementation of this technology requires making a number of corrections on raw scores to eliminate parasitic effects (altitude, presence of known neighbouring masses) and time drift of the apparatus (influence of the Sun and the Moon, tide *etc.*). But this first project on the Pyramid of Cheops had a particular attraction: it was the first time we used microgravity in a building, the mass surrounding the measurement apparatus. This constituted a challenging problem for engineers from both teams!

In April 1986, the EDF-CPGF team led by Jacques Montlucon and Pierre Deletie conducted the first measurement campaign inside the Pyramid.

Results: Microgravity found no misfit of density in the vicinity of the King's Chamber, from the floor to the ceiling. However, down the corridor of the Queen's Chamber, a defect density of the order of 30 microgals confirmed the existence of a cavity (microgal=μgal=10^{-8}m/s^2). Incidentally, outside and at the Southern foot of the pyramid, microgravity located the presence of a second "solar boat". Therefore, even if the main purposes of MM Dormion and Goidin had fund no clear experimental validation from microgravity measurements, it appeared likely that there was a storeroom down. It was then that Mr Guillemin returned to us by asking, on behalf of the Egyptian authorities, if we could perform an *endoscopy* (a medical like imaging technique to view the internal parts of the body) to check the existence of the cavity and see what it contained. We agreed to his request (Later, we bitterly regretted our decision), but the narrow corridor (105 cm wide and 117 high) did not have the necessary space to allow microdrilling (1 cm in diameter) in the desired direction. After three meters of drilling through a thick wall cladding of two Egyptian cubits (about 1 meter) thick, then through limestone and into a second wall, two Egyptian cubits thick, the drilling ended in a cavity filled with sand. This non-result created chagrin among the fifty or so journalists from around the world who "*wanted to see the Cheops's treasure through a micro-camera connected to the optical fibre inserted into the micro-drilling hole*". Voluntary leak of false information spread out without our knowledge, triggering a unpleasant media exposure and justified criticism by the Egyptology community.

What deductions could be made from the drilling results? The analysis of the sand taken from the cavity showed that it was not put there by accident. The sand is perfectly screened, but it is unclear what it was. The existence of a double wall confirmed the existence of a corridor or a room, or even an organized building, because the facing stones of inner corridors were again composed of perfectly adjusted blocks of 2 cubits thick.

Such were the results of the first projects.

The most interesting results for archaeology related undoubtedly to the following two projects that we undertook at the request of Dr Ahmed Kadry, President of the Egyptian Antiquities. The first related to the origin of cracks in granite beams in the King's Chamber ceiling and superposed granite beams above the ceiling, designed to deviate the heavy load of the construction above the King's Chamber. These huge rooms called "discharge chambers" in fact reflected the expertise of the architects of the Pharaoh, but their cracking beams worried Dr Kadry for reasons of elementary security. The results of these two studies were not known to the general public.

This became the subject of a detailed report for our sponsor, the Egyptian Antiquities Organisation and a presentation of the findings was made in December 1987 to the Congress on Archaeology in Cairo. Regarding the structure of the pyramid of the first study, the results were presented to a Congress of Geology in Athens in September, 1988 and a PhD at the University of Nancy written by Jacques Lakhmanan, entitled « *Traitement et Inversion des donnees gravimetriques : la micro-gravimetrie, son application aux recherches de vides* ». ["*Treatment and Inversion of data microgravity measurements and applications to seek out hollow volumes*"]

Two questions arose: The presence of cracks on the huge granite monoliths that form the discharge chambers reflects the action of a system of internal stresses around this room. In short, the pyramid had deformed under gravity loads. Do these cracks constitute a danger? What is their origin? To answer these two questions, we modelled the stress field inside the pyramid taking into account the variety of rocks and land foundation (to within 400 meters) and we used a high performance computer code for mechanical structures developed for studying the tunnels of the dams. The pyramid of Cheops was represented by a mesh of 2000 nodes, very dense around the empty interior and especially around the King's Chamber. The calculation showed that no known cavity or earthquake could explain the cracks. We then asked whether the existence of a cavity of as yet unknown volume and with a location around the King's chamber could explain the cracking. Just two positions, out of 40 that were tested, predicted a cavity of some volume, somewhere near the King's

Chamber that could cause the observed cracks in the observed locations. One was a small cavity underneath the collapsed Southern wall of the King's Chamber. The other was a cavity forming with the stone rafters (with very compact limestone) a "hard point" below the Northern wall. This result eliminates any danger to visitors, but raised questions for Egyptologists.

The study of the structure of the pyramid of Cheops was far more difficult. Its description and its results form the subject of this book, and are left for the reader to discover. As a result of these studies it has become clear that the Pyramid of Cheops is not a stack of homogeneous blocks. It is a relatively complex structure composed of granite, limestone and hard quarried Tourah limestone. Its density is not 2.7 T/m^3 (previously estimated on the basis of a homogeneous mass), but about 2.05. That this is a precise figure is shown in this book. Everyone can see that this magnificent building has resisted time, though somewhat undermined by human interference and abuse. It is perfectly organized, but in trying to understand its organization, we are like children in front of a black box that we cannot open. We must try to find out how it was made, and we can do this only from the outside. In this spirit, H.D. Bui, a researcher at EDF, who was already at that time a corresponding Member of the Academy of Sciences (1987), proposed a new method known since as the "inverse method". This method let him revealed an interesting aspect of the structure of the Cheops pyramid, namely the existence of interior ramps used in its construction, the starting point of a current and very attractive theory.

But you may ask, why wait more than 20 years to publish these results to the general public, except for the scientific papers presented at conferences in Cairo in 1987 and Athens in 1988? We have already given the bulk of the response. We were shocked and I would even say, insulted by the conduct of the first project on the unknown cavity. There was a clear order signed by the Department of Foreign Affairs of France but a total lack of a responsible partner *in situ*, no prior consultation with specialists Egyptologists, leaks to the press with misinformation suggesting that we were in fact looking for the Cheops's treasury, finally a swarm of reporters in the corridor to the Queen's Chamber!

The terms of our commitment to the following two projects were clear: No communication to the general public. If this policy was followed we would be able to work in peace. We set up a specific set of ethical guidelines for Technological and Scientific Sponsorship, we subsequently estimated that the results could hardly be of interest to the general public. In retrospect, I wonder if I neglected some essential information As the manager of these projects, had I misjudged their archaeological interest? It is true that at that time, I had to balance a number of roles, including Deputy Director of Studies and Research Division and Secretary of the Committee of Management Planning. Now I am inclined to think that the first written submissions which I set were overly cautious. Besides, four years later, on the basis of these documents, I wrote a book entitled « *Du Titanic à Karnak - L'aventure du Mecenat Technologique* » ["*From Titanic to Karnak - The adventure of the Technological sponsorship*"] Dunod, Paris (1994) and mentioned « *one keeps in mind that the top is less dense than the base and it is reinforced around the corners* ».

However, it seems now that I have not taken the time to discuss the results in depth directly with H.D. Bui because if I could share his presumption, namely that of an inscribed limestone ramp along the degree walls used in building the pyramid, I would unhesitatingly ask to produce additional data to validate and clarify this result.

It was tragic that Jacques Lakshmanan passed away shortly after, struck down by a serious illness. Also, we were approached by other projects in Egypt (Saqqarah, Luxor, then Tanis), and asked to treat objects out of the wreck of the Titanic.

So I personally closed the Cheops project files.

In 2000, the architect Jean-Pierre Houdin, examined the work of H.D. Bui and colleagues to see whether they supported his theory. He proposed an internal ramp tunnel, and from specific observations of the site, particularly in the Grand Gallery (which he considered to be a guide for counterweights for hoisting megaliths) he made an interesting explanatory model for the construction of the pyramid. He asked Dassault Systems to construct three dimensional models of various stages of construction, to further ensure the consistency of his model. Incidently, engineers from Dassault Systems reproduced the same type of calculation that we had performed to explain the cracking of the ceiling beams and came to similar conclusions. These studies of Jean-Pierre Houdin, supported by the Technological and Scientific Sponsorship of Dassault Systems (from which we are very happy to benefit!) were published in 2006 in a book entitled "*Khufu, The secret behind the building of the Great pyramid*" Farid Atiya Press. They were then shown and broadcast in Television in 2009. Although several elements of the model by Jean-Pierre Houdin require clarification, especially near the top of the Pyramid, nobody can be indifferent to the fact that the stones of the outer coating were adjusted so well on all sides (with a slight inclination inward to the surface, as we can observe in the Meidoum Pyramid). Houdin's model has the merit of coherence, which is essential for architecture and has to be confirmed or invalidated.

We realized that now is the time to publish our definitive contribution to the scientific debate, giving all of our information, including some that had remained unpublished for twenty years.

By doing so, we have achieved the main purpose of the book.

Marc Albouy

Introduction

This book tells a story of an encounter between Archaeology and Applied Mathematics. It is devoted to the imaging of the surface density distribution in the Cheops pyramid. It concerns the results on the surface mean density obtained by our team working in 1986-1987 on the Cheops pyramid project, using experimental data on the gravity measurements at the Cheops pyramid site and mathematical & numerical methods to solve an inverse problem. The mean density corresponds to the density of finite elements along the external surface of the pyramid. The study was done under the Technological and Scientific Sponsorship of Electricite de France (EDF): *a sponsorship unlike others*, according to its director Marc Albouy, in his book (1994). In an ordinary sponsorship, the granting agency gives money to support a project. Here, EDF gave not only money for gravity measurements at the Giza site done by CPGF (a French Company specialized in Geophysical Exploration), but also the know-how of EDF Engineers and Researchers, including me. EDF also provided computer facilities to work on the project (Super computer CRAY 1 at its Computer Center, at Clamart, *etc.*).

This book tells the story of two microgravity operations on the Cheops pyramid, the first one searching for an unknown tomb of the Pharaoh which failed in autumn 1986 and the second one devoted to the entire pyramid carried out after the failure of the Cheops project, without EDF's support or knowledge. The result of the second operation was published in the Proceedings of the Athens Symposium on Geology, P.G. Marinos & G.C. Koukis (Eds), A.A. Balkema (1988), pp. 1063–1069. The premature ending of the Cheops pyramid sponsorship prohibited us from any complementary gravity data or more refined computations or graphical displays.

I never published the image of the density distribution, except on my web site (http://hdbui.blogspot.com). Its main conclusions were reported in the Athens Symposium Proceedings, but it was not published in our paper of the Proceedings because of a lack of time for submitting our paper. The image was ignored for 12 years until July 2000, when my colleague Pierre Deletie, an EDF geologist, showed it to Henri Houdin (Engineer) and his son Jean-Pierre Houdin (Architect) who were working on a new theory regarding the Cheops pyramid construction.

Chapter 1 presents some microgravity studies at EDF. We will discuss some technical points on the Cheops pyramid which are not known, and the mystery of the Cheops's funeral chamber.

Chapter 2 is devoted to Microgravity as a Geotechnical method used in Engineering. It explains the difference between *Exploration* and *Inversion*. It mentions the *blind test* which was done for the purpose of convincing CPGF to work with us on the Cheops project.

Chapter 3 presents the results on the density imaging of the pyramid.

Chapter 4 makes a tentative interpretation of the density image in a *Virtual Reconstruction* of the pyramid. We do not introduce a new theory of the Cheops pyramid construction, but consider only *existing* theories, some of them are consistent with our density results. Two theories are found to be supported by our densitogram. This illustrates what is already known by Mathematicians: an inverse problem may have more than one solution.

Chapter 5 gives the details for a graphical interpretation of the results.

We thank Jean-Pierre Houdin for permitting us to use some of his photos (http://www.construire-la-grande-Pyramide.fr). I thank Marc Albouy, my former Director in charge of the Technological and Scientific Sponsorship of EDF, for writing the preface and for inviting me in 1986 to participate in his Cheops Pyramid Project, which finally gave a cultural and archaeological dimension to our mathematical, mechanical and numerical works. I wish to acknowledge my co-authors of the Athens Symposium paper (1988), J. Montlucon, J.C. Erling and C. Nakhla for their contribution to Chapters 2 and 3 as well as M. Bonnet and X. Chateau from LMS/Ecole Polytechnique, Yves Lemoine and Jean-Pierre Baron from CPGF, Jean-Pierre Lefebvre and Yves Wadier from AMA/R&D/EDF. My thoughts are with Jacques Lakshmanan and Pierre Deletie who passed away ten years ago, with whom I shared a passion for Egyptian Antiquities. I express my admiration to those who conceived the Cheops pyramid and to those who built it so long ago.

I would like to dedicate this book to my family Marie, Raphael, Cathy & Jean, Joanna, Oriane and Theo who provided me the photo of the Cheops pyramid.

Last but not least, I warmly thank Alan Rodney for his careful reading of the manuscript and for many interesting suggestions.

Palaiseau, France H.D. Bui

Contents

1 **On the Cheops Pyramid Studies** 1
Historical Context of the Studies 1
The Mystery of the Unknown Chamber 4
What We Know and Do Not Know in the Pyramid? 5
The Petrie Sequence and the Puzzle of Stones 7
Herodotus and Sauneron 10

2 **Microgravimetry in Geomechanics** 13
A High Precision Balance 14
Exploration of Sites 15
The Limitations of Exploration 16
Inverse Problem and the Butterfly Effect 17
The Working Conditions in the Cheops Operation 19
The Blind Test 21

3 **Density Images by Microgravity** 25
The Second Solar Boat Discovery 26
The Measurement Campaign in the Pyramid Site 27
Measurement Results Near the King's Chamber Structure 27
The Low Mean Density 2.05 T/m^3 of the Pyramid 29
Interstices and Voids 31
Direct Computation of Gravity due to a Cavity 33
Inversion of Gravity Data for Cavities Near the Chambers 35
Some Mathematics of the Inversion 36
Imaging the Pyramid with Microgravity Measurement 39
Three-Dimensional Meshes of the Pyramid 41
Results on the Imaging of the Surface Density 43
The Densitogram 46
Raising the Density 48

4 **Virtual Reconstruction of the Pyramid** 51
The Holscher Ramps and the Steps of the Construction 53
Macroscopic and Microscopic Points of View 57
The Densitogram and the Borchardt Pyramid 58

The Houdin Internal Ramp Tunnel 60
The Mystery of the King Tomb 62
Golden Number and Intertwined Spirals 63

5 Filling the Cornices 67
True Density and Mean Density 67
Comparisons with Observations 69

Notes 71

Bibliography 77

Permissions and Acknowledgements 79

Index 81

About the Author

H.D. Bui graduated from Ecole Polytechnique, Paris and the University of Paris VI. He had an original professional path. He first worked with EDF (French Electricity Company) in 1961 on problems of Solid Mechanics related to the National nuclear energy program, then in 1982, took part in the creation of the Department of Mechanical and Numerical Modeling at R&D Division, EDF. Throughout this period and even until now, he worked in two places Electricite de France and Ecole Polytechnique, where for about fifteen years he was associate Professor of Mechanics. He is currently a research director of Ecole Polytechnique ParisTech, Laboratory of Solid Mechanics LMS (UMR X-CNRS 7649), a researcher at the Laboratory of Durable Industrial Structures (UMR EDF-CNRS 2832) and at the Mechanical Unit of ENSTA ParisTech.

Since his studies on the pyramid of Cheops in 1986, Inverse Problems have become his favorite subject. He is internationally known by his works in this field, as indicated by his presence until 2005 on the Editorial boards of Mathematical Journals: *Inverse Problems in Engineering Sciences (Canada), Inverse Problems (USA), Inverse and Ill-posed Problems (Russia)*. His first book *Inverse Problems in the Mechanics of Materials, an Introduction*, in French, Eyrolles Paris 1992, was translated into many languages (English, Japanese, Russian, Chinese). His latest book *Fracture Mechanics, Inverse problems and solutions*, Springer (2006) has been translated into Russian, Fizmalit (2011).

His two research teams at EDF and Ecole Polytechnique have solved major inverse problems in Acoustics, Thermal conduction, Elasticity, Thermoelasticity, Dynamic Viscoelasticity, Elastodynamics with applications to Non Destructive Evaluation of materials, Earthquake inverse problems and Medical Imaging.

H.D. Bui is Member of the French Academy of Sciences, the European Academy of Sciences and Fellow of the French Academy of Technologies.

Chapter 1
On the Cheops Pyramid Studies

We can see a short distance ahead but we see plenty there that needs to be done.
A. Turing

Historical Context of the Studies

One afternoon in May 1986, working as Deputy Chief of the Department MNM (Mechanical and Numerical Modelings, EDF) I had a visit of my Director Marc Albouy, in charge of the Technological and Scientific Sponsorship of EDF. He wanted my help to provide an answer to some questions addressed to him by the French Embassy at Cairo, on behalf of Egyptian Antiquities Department : « *Does EDF have some means for detecting cavities in the Cheops pyramid* ?". Of course, the answer was the microgravity technique used in our operations to detect cavities and defects under future hydroelectric dam sites and power generation plants. The question from the highest authorities of two countries was worthy of our attention and at the same time embarrassed us, because it was a new kind of problem which we had never studied, to detect an unknown cavity itself and not only to detect a gravitational anomaly which signalled the possible presence of a cavity. As will be seen later, the difference between detection of a gravitational anomaly and detection of a cavity is as large as the difference between *Exploration* and *Inversion* of the gravity equation.

In the Sixties we had worked on Mechanical problems for the construction and maintenance operations of dams for Hydraulic energy. Since then, new problems arising in the Nuclear Energy program were prominently in our daily works. Problem fields ranged from Fluid Mechanics, Thermal Conduction, Electromagnetism, Neutronics, Solid Mechanics and also Mathematics and Computer Sciences. Numerical analyses and Mathematical methods were of great importance for our Company, for both design problems and maintenance problems.

One interesting problems among others was the detection of cracks and defects in structures, which could be proved dangerous if not discovered before going into

H.D. Bui, *Imaging the Cheops Pyramid*, Solid Mechanics and Its Applications 182,
DOI 10.1007/978-94-007-2657-4_1, © Springer Science+Business Media B.V. 2012

service. New mathematical *Inverse Problems* arose then. They were, roughly, problems in which one searches the *causes* from the knowledge of the *effects*. We notice that a Forward problem deals with the reverse. Mostly, up till then, we were looking for possible effects from the knowledge of the causes. Here is an example of inverse problem. In the so-called Leak-Before-Break (LBB) problem, one can measure the leak rate of pressurized and hot water coming out of a longitudinal crack in a pipe and then determine the crack length. It is not a simple inverse problem since the crack is hidden behind a thick insulator. A well known proverb gives another example of inverse problems " *Tell me who your friends are and I'll tell you who you are*". Another familiar example is the diagnosis of our Doctor who uses a stethoscope to listen to our chest. You might also remember Henri Vincenot, the French author of the book « *Memoire d'un enfant du rail* », [*The memory of the child of a railwayman*] and the railwayman who slaps on the wheels and listens to the abnormal sound when there are defects?

They are inverse problems and the problem raised by the Egyptian Antiquities who wanted a non destructive method of investigation on the unknown tomb of the Pharaoh is one of these. The gravity method requires two distinct operations led by two different EDF teams, one experimental (measurement of gravity around the pyramid site) and another theoretical one led by the author (mathematical & numerical methods of inversion of the Newton gravity equation).

What surprised us was the nature of works to be done by our theoretical team at the R&D Division, who was not familiar with archaeological studies. But we had no choice to refuse to work on such a prestigious site! There were two successive works on the Cheops pyramid. Just before the visit of Marc Albouy, another EDF team solved a mechanical problem for two architects, G. Dormion and J.P. Goidin, who made the conjecture that there might be an unknown chamber near the current King's Chamber, next to the Great Gallery, basing on their observation of stones of the North wall. The aim was also to understand why there were cracks in the granite ceiling of the King's Chamber, in the intrados and extrados of the granite beams [(1)]. A relative vertical displacement of the Southern wall was observed. Was this due to the presence of a cavity next to the Northern side or below the current Chamber? Perhaps was it the unknown tomb of the Pharaoh? The numerical analysis by a 2D finite element method modelling, justified by the longer side of the Chamber in the East-West direction, was performed by my colleagues Jean-Pierre Lefebvre and Yves Wadier, with the aid of Pierre Deletie for the choice of material constants, taken from the data base of stones, having the same geological aspects and characteristics of the Giza site (Tourah stones) or the Assouan site (Granite). The Finite Element computation results were never published, except the EDF report [*Etude geomecanique de la chambre du Roi dans la pyramide de Kheops*», J. Montluçon, J. P. Lefebvre, Y. Wadier, T. Lapointe, P. Deletie et P. Martinet (1986)]. Following this report, it was discovered that there were indeed two cavities below the King's Chamber which could explain the ceiling cracks. Recently a team from Dassault Systems (France), working for Jean-Pierre Houdin (2009), reconfirmed our colleagues' apostrophe results. Cracks resulted indeed from the differential vertical displacement of the Southern wall, but Dassault Systems did not investigate further on the origin of this displacement.

The second study was requested by French and Egyptian authorities, for a micro-gravity detection of the unknown tomb, using techniques which were well controlled by EDF in the Exploration of dams and tunnels, with the technical aids of CPGF (the French Company of Geophysical Exploration) who worked with EDF a long time on dam exploration. However, EDF wanted to go further than a simple exploration of the site, trying a true inversion of the Newton gravity equation in order to recover the density distribution. A cavity zone might correspond to the null density in the solution. This inverse problem was as delicate as trying to find *a needle in a haystack* said my CPGF colleague, Jacques Lakshmanan. Preliminary results showed that there was no significant anomaly near the Northern wall of the King's Chamber, but only some indication of an anomaly midway along the horizontal corridor to the Queen's Chamber. Finally, the third study, while requested by the Egyptian Authorities, was not supported by EDF after the failure of the second study, about which we will talk later. My colleagues and I decided to pursue the mathematical & numerical analyses on the *whole* pyramid, using data gathered in the second study, at a scale that ignored the internal unknown cavity near the King's Chamber. The results of the third study are reported in this book. Contrarily to the second study about the unknown cavity near the King's Chamber, which is not yet an inverse problem, we are dealing here with a true inverse problem. We must remind ourselves that *exploration* to detect anomalies is comparable to the diagnosis of our doctor who examines our chest with a stethoscope. Then, using his experience, he prescribes the medicine as appropriate. It is not comparable to modern apparatus in hospitals, where specialists make use of Radiography, Scanners, Acoustic Scattering, or Magnetic Resonance Imaging (MRI) to "*see*" our body in the inside. One does not always realize that these apparatus used in Medical Imaging were invented after mathematical studies and discoveries, generally unknown to the public, for example Radon's transform at the beginning of 20th century and its inverse transform used in the Scanner technology, the *attenuated* Radon's transform recently discovered by Novikov (University of Nantes) and more recently the *conical* Radon's transform by two researchers, Nguyen and Truong (University of Cergy-Pontoise, Eastern Paris). The last transform has been studied for powerful Gamma ray scattering, which can penetrate a plate of steel 40 cm thick so why not the entire Cheops pyramid as well? [See Bui, 2006].

In short, we were asked to realize an *Imaging of the Cheops pyramid density*, even for the second operation on the unknown King's tomb, which we never done for our own dam sites. As said before, to solve an inverse problem we need data and data processing. In the case of a X ray scanner, data processing is very simple with the Radon's inverse transform (no attenuation). For gravity inverse problems, no such a simple inverse formula exists and we need a computerized inversion method. Our associate, CPGF, in charge of microgravity data for EDF, handled the general exploration, looking at a graphical display of data to draw conclusions as to the presence of an anomaly or not, occasionally with a good result in the case of data collected on planar site, for example in the case of the discovery of the second Solar boat at the Southern side of the pyramid, but not in the three-dimensional case. There was a need for a confirmation by drilling a hole. Unfortunately, it seems impossible to drill a hole in the Cheops pyramid, because « *it would be like drilling a hole in the heart of Egypt* " said Zahi Hawass, President of the Egyptian Antiquities.

In 1986, based on the exploration result along the corridor to the Queen's Chamber which suggested a possible presence of a cavity, not waiting for the results of our theoretical works, about mathematical and numerical inversion of the gravity equation, that were still in progress, the experimental team obtained the necessary Egyptian authorization for drilling three inclined holes, downward in the East direction. Encouraged by the success of the second Solar boat discovery by a simple exploration, this experimental team was anxiously waiting for the media success of the discovery of the Pharaoh treasure. In the case of success, the team leaders should be renowned like Champollion who deciphered the Rosetta stone (discovered by P.F.X Bouchard of Ecole Polytechnique) or the discovers of the Touthakhamon treasures. Unfortunately, it was not the case. The drillings found nothing else but some traces of yellow sands between stone walls. After the media failure of the second operation which signalled the end of the whole Cheops operation, we might stop our works on the Cheops pyramid.

Later, we found the reasons of the failure which could be explained first by the precipitation to be the first to discover the treasure. After all, the unknown cavity could very well exist, but not at the location wrongly indicated by the exploration result. The other failure reasons are possibly: insufficient data, mismatch between data and unknowns, inaccuracy of the inversion. In short, the main reason of the failure, at that time, was that exploration was confused with inversion.

In 1987, Jacques Lakshmanan (CPGF) and I, decided to continue our work on the gravity inversion, without the support of our respective companies, without money for CPGF since data were already available and without the knowledge of EDF, this time for studying the whole pyramid. The number of data was not sufficient enough to detect a cavity of about 125 m^3 inside the whole Pyramid, we shall divide the pyramid into coarser finite elements and look only at the overall structure. What did it result from the third operation? Perhaps one paper more in the Athens Symposium 1988? But most important was for us the feeling of having participated in the fantastic operation on the Cheops pyramid to pierce its mystery.

The Mystery of the Unknown Chamber

The granite sarcophagus of the King Chamber was discovered empty, presenting traces of burglary in a corner, the cover disappeared. Obviously, the profaners wanted to open the sarcophagus blocked by an extraordinary system of closing which showed already the perfection and the ingeniousness of the Egyptians, supervised by the famous architect Hemiounou. But the know-how of the Egyptians was not limited to this detail. The construction of the pyramid itself constitutes another great mystery for generations of Egyptologists who considered this question and for a long time, as long as one still does not find documents, traces of papyri, graffiti on the way in which the pyramid was built.

But let us return to the current King Chamber which one can visit today. It is for the ones only one lure, or the others more probably the tomb of Ka, *i.e.* the double of the King. This duality King-Ka came from an old tradition following which the King reigning like the suzerain of the two parts of the country, the High Egypt and the Low Egypt, governed each part by a King, was buried in two distinct tombs, one in the North containing his body and the other in the South, which was rather a symbolic tomb, called cenotaph, to honour his double. There are tombs of the Kings at Saqqarah in the North, at Memphis and tombs of their doubles in the South at Abydos. Cheops perhaps changed this tradition of tombs separated into two different places but preserved the duality King-Ka.

It is possible that the King and his double are always in the same pyramid. That can explain the construction of several successive tombs, initially the underground, abandoned room because of cracks in the ceiling during the digging and perhaps more simply, like it have been advanced by Egyptologists, for lack of ventilation for the workmen, then the Queen's Chamber and finally the current King's chamber, who would be abandoned in his turn because of cracks discovered in the ceiling granite beams. Some authors, Lheureux and Marin (2008), considered that cracking appeared much later and the King's Chamber was a hydraulic machine to move a secret door to the unknown tomb, while ventilation conducts would serve as the conducts to fill the machine with water. But where then the true unknown room of the King is?

It is the greatest mystery of the Cheops pyramid. Dormion (2004) working on the anomalies observed on the flagstones of the Queen's Chamber thought that Cheops envisaged another room in the West on a level a little below with access through a corridor to the East. Other archaeologists like Bardot and Darmon (2006) located it between the King's Chamber and the Queen's Chamber while basing themselves on the geo-mechanical work of the first EDF operation (J.P. Lefebvre and Y. Wadier's report) and especially on their discovery of true "forgery joints" stones of the wall of the corridor to the Queen's Chamber.

One can envisage another simple hypothesis. At different steps of the construction, it seemed likely that a temporary tomb might be *always* ready for use, a thick granite open tomb *without* door, with access only to the storage room, deeply inside the pyramid. At a higher level of the construction, another temporary open tomb would be prepared while the unused former one below was filled of stone and sand *etc*. The inviolate tomb would have a corbelling granite roof and absolutely no access to the exterior!

What We Know and Do Not Know in the Pyramid?

The examination with the naked eye of stones in all the parts accessible from the pyramid makes it possible to have a good idea of materials used. There are several kinds of stones, limestone of the Giza area for the main work, which we call filling stones, with density about d=2.35 (T/m^3), Tourah stones of the walls, or simply limestone of coating used in the interior corridors, granite wall, granite ceilings and beams of the discharge chambers, of density about d=2.65 T/m^3. The top of the rooms of discharge consists of two Tourah limestone beams of the rafters, while supports of the granite beams are made of limestone. The ground of the King's Chamber is made of thick flagstones.

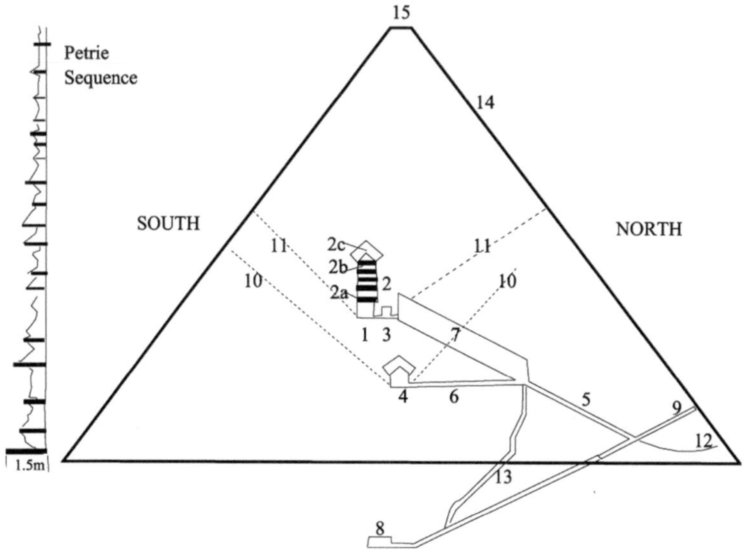

1. King's Chamber
2. Discharge Chamber, 2a Granite beam
   2b Chamber, 2c Chevron
3. Harrow chamber
4. Queen's Chamber
5. Ascending corridor
6. Horizontal corridor
7. Grand gallery
8. Underground chamber
9. Ascending corridor
10. Ventilation duct
11. Ventilation duct
12. Al-Mamoun hole
13. Exit channel
14. Limestone coating
15. Small top pyramid (if existed)

Fig. 1.1 North-South section of the pyramid and Petrie sequence. In the left, the Petrie sequence shows the thickness versus the height; maximum thickness 1.5 m at the bottom, 1.25 m at the 19th layer. The Petrie sequence of thick stones is not correlated to the main structure. They would be linked to the stability of internal degrees, served as the control of horizontal levels and the reference marks to carry out the final geometry. The pyramid has a square base of side 230 m, a height about 146.50 m

Except visible stones on internal walls and ceilings, one does not have an idea on stones inside the pyramid. What is the mean density of the pyramid? Does the Cheops pyramid consist of a system of degrees like many other pyramids? What about stones near the surface of the pyramid? The white stone coatings disappeared long time ago and current visible stones are limestone, more or less well arranged.

The Petrie Sequence and the Puzzle of Stones

Seen outside, visible stones, which we call *facade stones*, are of the second or the third rank behind the disappeared finishing stones. W.F. Petrie (1880) recorded the thicknesses of facade stones going up to 1.50 m for the 19th rank and obtained what is called the *Petrie sequence*. Apparently there is no of particular nature in the succession of thicknesses. There is no either obvious correlation with the structures of the pyramid core. This raises some interesting unsolved questions.

Do stone bases of strong thicknesses of the Petrie sequence constitute the foundations of higher degrees masonries? From mechanical and technical point of views, it is highly probable that one cannot build masonries of the degrees on non-arranged and deformable ensemble of stones like a granular material.

Layers of thick stones of the Petrie sequence have some interests:

1. From time to time to update the horizontal level of bases, after having posed stone bases of less thickness and to put reference marks to carry out the final geometry,
2. to have a solid foundation raft to continue the masonry of the degree wall,
3. to ensure the stability of the work and to have a solid connection between the degree walls and the stone bases, under the *condition* that the stone base is undeformable as a whole. This condition can be satisfied by two methods: the first one using mortars between base stones, the second one by considering each stone base as a *puzzle of stones* without mortars. We know that puzzle for kids is rigid in its plane because of the complementary geometry of each puzzle piece.

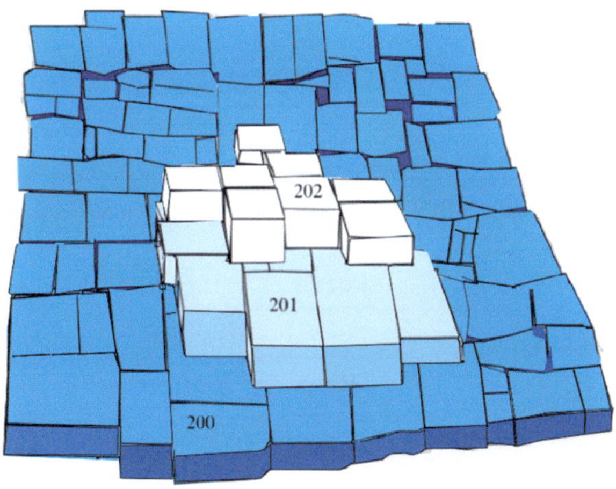

Fig. 1.2 The puzzle stones of the top 200th layer (*Blue*) and the remnants of 201th (*light Blue*) and 202th (*White*) layers are not put down in quincunx with straight lines through the layer, but rather like a puzzle of variously shaped and complementary stones, without mortar (Approximate drawing)

Fig. 1.3 The low Temple of Khephren. Puzzle stones in the wall (Sebi, Wikimedia)

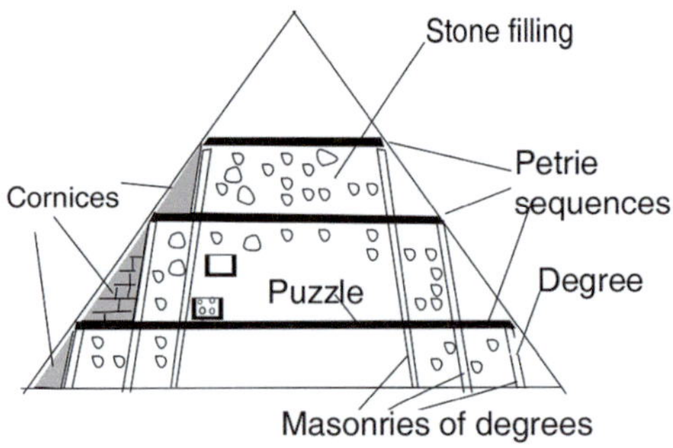

Fig. 1.4 The assumed geometry of puzzle stones of the Petrie sequence, as the foundations of degree wall masonries. The spaces between the Petrie sequence and the degree walls are filled with stones of various shapes, with or without mortars, comparable to the ones which can be seen through the big hole of the Meidoum pyramid wall. The cornices are filled with squared stones with more or less interstices or gaps. The square U shaped drawings at different height are temporary open granite tombs, filled with stones or sand while unused

If one sees the aspect of the stone arrangement of the 200th and 201th summit platforms, to approximately 136 m, one has the feeling that each layer looks like a puzzle of complementary shape stones, making stones interdependent from/to each other, like only one rigid block, a base of strong cohesion in its plane, Fig. 1.2. There are no straight joint lines across the whole top platform which would create weaknesses or slip lines. Did bases of strong thickness extend horizontally on the whole inner pyramid? Were the stones of strong thickness put down like puzzles likely to rigidify bases? The answers are not known. One is struck by the virtuosity and the precision of the builders when one looks at the walls of the low temple of Khephren and sees these enormous stones placed as *puzzle stones* to prevent the horizontal slip by an eventual earthquake. In small constructions of houses nowadays, the two processes exist: a foundation raft extended under all the house or a foundation of width 3 to 4 times that of the walls. In the same way, the bases of strong thickness assembled in stone puzzle and extended on the whole upper surface of the degree, except the core, would reinforce the building. One would have thus built a solid and stable ensemble of stones *without* any mortar plasters.

This question of stone bases of strong thicknesses is relevant when one thinks of the ruin of the pyramid of Meidoum. One sees there a smooth frontage of an old degree of 32 m height on which the reported degrees had slipped down probably after an earthquake, cf. [Kerisel, 1991].

If there were stones of strong thickness, anchored firmly in the core of the old degree, there would be no slip of the additional blocks, because of a larger cohesion between old and new ones. But the architects of Snefrou, the Cheops's father, did not have this possibility since it was about the transformation of an old pyramid by external additions.

Fig. 1.5 Meidoum pyramid. Through a big hole in the wall at the left, one can see the thickness of the wall masonry and small blocks of filling stones of various shapes (Permission of J. P. Houdin)

The reinforcements of stone blocks are known to exist in Archaeology and Engineering. Kerisel mentioned in his book the reinforcements of the stones of the Newgrange cairn, dated 3200 BC, by grass sods. Among other examples of reinforcements, there are internal masonry walls (Senostris II pyramid), trunks of palm trees in granular stones, steel cables in the slopes of highways, recycling worn tires in the floor of roads according to the *Pneusol* process by Nguyen Thanh Long (1985). For his pyramid the architect Hemiounou, would have the idea of reinforcement by stone bases of high thicknesses, the Petrie sequence, we can rather say the *Hemiounou sequence,* which has extremely well resisted for 4500 years the bad weather and earthquakes.

Herodotus and Sauneron

Let us finish this brief review to evoke an aspect of the History. The Cheops pyramid was built about 2530 BC. Two thousand years after, in his *Relation of voyage in Egypt*, Herodotus evoked its construction such as it was told to him. Certainly, there are elements of truth, as there is the doubtful one or approximate one in its account. We will return on what he told, in connection with the construction itself, in the following chapters. What about the number of workmen and their life & working conditions? Nothing very sure, if not an example quoted by G. Goyon (1977) of a forwarding to the careers organized by Wadi Hammamat which brought together 18 000 people. It was organized like a true army. Can one accept the account of Herodotus which spoke slaves?

It appears difficult it to us to believe when one sees the perfection of the construction of the pyramid carried out by people which should really accept so that it made, more especially as a text of Sauneron, quoted by G. Goyon at Ramses time II (1200 BC) quite former to Herodotus, who spoke in the paternalist way in which the Kings dealt with their people who built their pyramid "*I ensured your subsistence in all produced, thinking that you would work for me of a grateful heart* ". Goyon has certainly reasons to think that this text can also apply to the period of Cheops. We also think it. Recent discoveries by Z. Hawass of Egypt left no doubt about this subject. He discovered the tombs of workmen very near from Royal tombs, which indicated a high recognition of their status.

What is of Cheops himself known? Not large things of precise, if not which he succeeded his father Snefrou who built the pyramid of Dahchour and Meïdoum towards 2550 BC. Also remains of him the drawing of his name and especially a small statuette out of ivory of 7.5 cm (Cairo Museum). His face reflects the majesty, serenity and the mystery which we will find in his pyramid.

Fig. 1.6 Statuette of Cheops out of ivory of 7.5 cm (Cairo Museum)

Chapter 2
Micrograviметry in Geomechanics

"Nil novi sub sole" [Nothing new under the Sun]
Salomon, in Ecclesiate I,10

The gravity which we feel everyday is still the most mysterious force of Nature. Even today, physicists do not know how to explain it, how to unify it with the three other forces, *viz*, electro-magnetic, nuclear and weak forces of the Standard Model of Physics. They do not how know is the exact nature of the *graviton* which carries the force of gravity in analogy with the photon which carries out the electromagnetic force [(2)]. Here we consider simply the ordinary gravity force known from Newton's law, according to which two masses m and m' at distance r, are instantaneously and mutually subject to the attractive force in the now well-known inverse square law

$$F = k\frac{m\,m'}{r^2}$$

k=6.6726±0.0005 10^{-11} kg^{-1} m^3 s^{-2} (universal constant). The Earth, considered as a lightly flattened spherical body, geometrically speaking, an oblate spheroid, of radius R=6 368 km and mass m'=5.98 10^{24} kg, attracts any mass m by the force F=mg. This formula defines the vertical gravity g felt by the mass m. Using the unit of gravity introduced by Geophysicists *i.e.* the Gal, the gravity on the ground is about g=980 Gal. With the unit μGal (10^{-8} m s^{-2}), the gravity on the ground is about g=9.8 10^8 μGal. One uses this smaller unit μGal for detecting negligible variations of gravity due to many causes: variation with the height, time dependent tide (influence of the Sun and the Moon), presence of important mass of the mining deposits under the ground, or presence of an underground cavity *etc.*

On the planned zone for the building of a dam or a EDF nuclear power plant, or for the search of an underground cavity inside a salt mine area, one is not interested in the absolute value of gravity, but only in the variation of gravity when one moves on the zone studied in a radius of 1 km. The measured gravity variation is negative $\Delta g<0$ when there is a cavity below the measurement apparatus and positive $\Delta g>0$ when there is a dense rock formation, for example a granite, which is heavier than the surroundings. One assumes that, in the absence of gravity anomalies, the measured gravity should vary slightly, or within some limits according

H.D. Bui, *Imaging the Cheops Pyramid*, Solid Mechanics and Its Applications 182, DOI 10.1007/978-94-007-2657-4_2, © Springer Science+Business Media B.V. 2012 to a law called the *Bouguer correction*. This correction is only valid inside some small zone according to either an affine law $g_B=g_0+ax+by+cz$, or a quadratic one $g_B=g_0+ax+by+cx^2+dy^2+fz+ez^2$, where the constants can be adjusted to measurements, or considered as unknowns as well. This will be used in our study of the Cheops pyramid. For a larger zone about 100 m to 1 km, one is considering the *Bouguer anomaly*, which is characterized by gravity level sets, which reveal the natural anomaly of a heterogeneous area. The search of sand deposits for civil engineering materials makes use of the microgravity techniques. As another application of microgravity we can mention the precise information and verification on the density distribution obtained in the drilling operation of the basement and lower levels. Another well known example is provided by the discovery of chromium ore deposits in Russia, using surface measurement of microgravity, Fig. 2.1.

A High Precision Balance

In Geological earthworks, one uses a measuring device which is a kind of balance of utmost precision equipped with electronics, called balance. The principle of the measurement apparatus is rather simple. A mass m is fixed at a spring. This mass is more or less heavy according to the gravity of the place (latitude and altitude). If the measuring device moves to another point, the gravity changes from g to $g+\Delta g$, what causes a variation of the weight $\Delta P=m\Delta g$ and the additional elongation of the spring. By electronic devices, one captures the elongation of the spring and hence the gravity change. The sensitivity of the Lacoste & Romberg gravimeter [(3)] is such that the variation -30.8 μGal is detected when the balance is raised up to only 10 cm. This is simply because the centre of the Earth moves away 10 cm and thus attracts the mass m less.

Since the Earth is a flattened geoid, the gradient of gravity $\partial g/\partial x_3$ which is measured with precision, is found to vary with the *latitude*, from 308.779 μGal/m (latitude 0°) to 308.338 (latitude 90°). The increase (or decrease) of the atmosphere pressure changes the mass of the air above the balance, according to the gradient $\Delta g/\Delta p=-0.36$ μ Gal/mbar (or -3.6 μGal/KPa). Here, a higher pressure increases the air density and the gravity decreases because the additional mass of air above the balance attracts the mass m upwards. Another influence to be considered is the *tide* in zones near the sea, such as the power plants in Normandy (Gravelines, Flamanville, Penly). The variation of gravity in Normandy is about $\Delta g/\Delta h=0.02$ mGal/m. In the Cheops pyramid region, we took account of the influence of the Sun and the Moon which depends on the date and the horary, provided by a microcomputer. Each gravity measurement is performed at various time intervals of some minutes. Every hour, or at the beginning and the end of the operations, we check the reference basis of gravity.

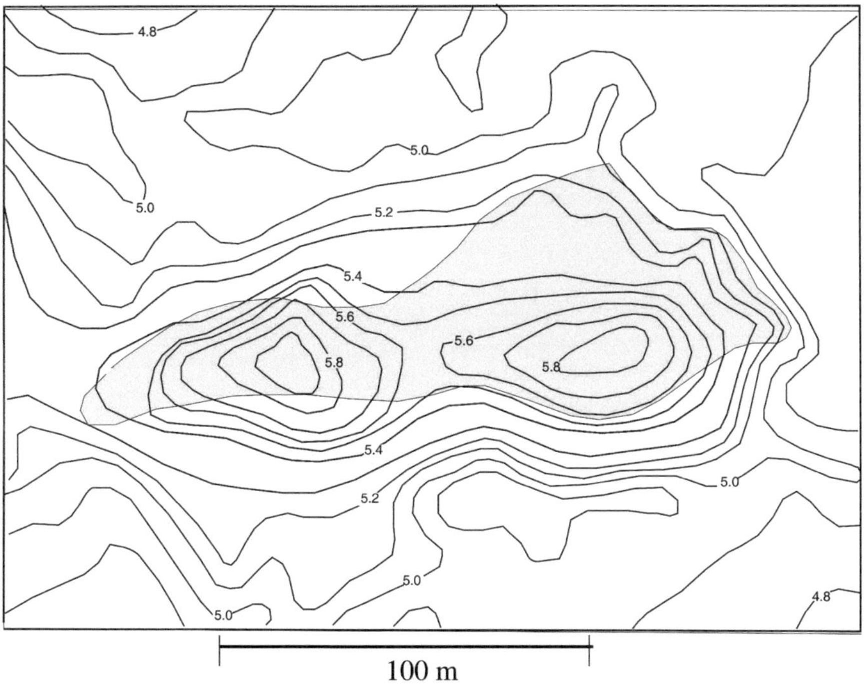

Fig. 2.1 Chromite deposit in Russia. Level sets of gravity (mGals) After Mironov and Lashkmanan's PhD thesis

Exploration of Sites

The use of a high precision balance is the simplest non destructive method to explore the future site of construction of dam or power plants, or for controlling the cavity used for stocking gas in a salt mine, or for detecting cavities in a railway embankment. The implementation of the microgravity method is very easy with a portable balance. The real cost of the operation is that of skilled labour qualified to take measurements and to carry out analyses of the experimental results. The total cost is much less than the cost of a destructive drilling.

To end this brief discussion on gravity applications, we mention two other examples: the search of gravity anomalies in the site of the Coche dam (EDF), a small 34m high structure, built in 1972-1975, above the Savoie town of Moutiers at an altitude of 1 401 m and the Petra-Marina site (Corsica) for ensuring that the underground of a future building is free of buried explosives left since World War II.

In 1986, two political and cultural authorities of France and Egypt in Cairo addressed us the following question: "*Does EDF have a means for detecting a cavity inside the Cheops pyramid?*"

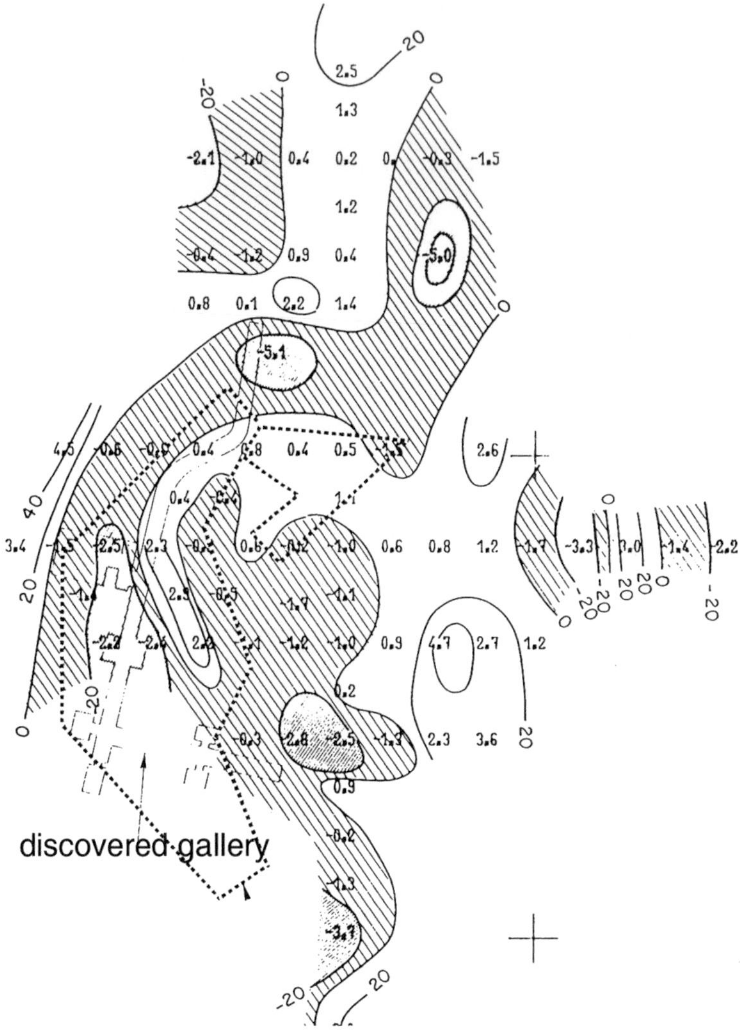

Fig. 2.2 La Pietra Marina (J. Lakshmanan PhD Thesis, Univ. of Nancy 1990)

This is how Micrograyity exploration is came to be used for the first time to study the Cheops pyramid.

The Limitations of Exploration

One knows that *exploration* is a qualitative method of inspection to detect the presence of anomalies in underground. It is similar to an examination of our body by our doctor who uses the stethoscope to listen to our chest. One performs gravity measurements without any inversion of the data, not to recover the density distribution, but to establish the level sets of gravity.

A counterexample is given in Fig. 2.3 where the deficit of gravity is theoretically given for two cases: a cavity filled with a gas and a cavity filled with brine, with different radii and depths. The anomalies are found to be very similar in both cases so that it is difficult to know the exact geometry of the cavities (diameter and depth) and the density of the filling. To get some differences, one has to display level sets of gravity corresponding to different altitudes x_3. However, the presence of two cavities is well indicated by level sets of gravity.

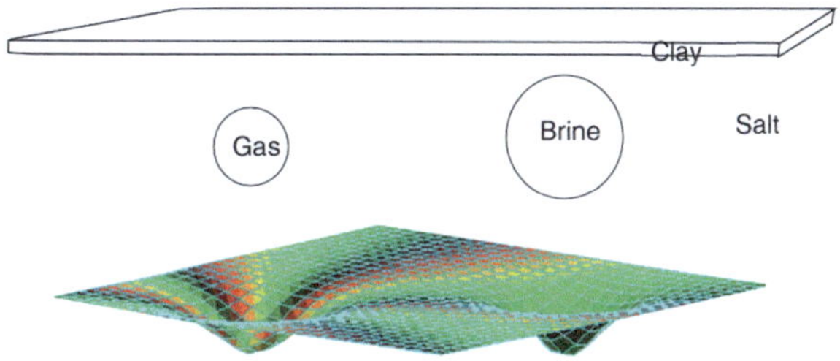

Fig. 2.3 Salt mining recovered by a clay layer with two spherical cavities filled with a gas and a brine. Theoretical gravity at the ground level for: (i) Gas (radius 4 m, depth 5 m, density 0); (ii) Brine (radius 5, depth 6, density 0.8)

Inverse Problem and the Butterfly Effect

For those who already have experience of the exploration of sites, they will find that the deficit of density is smaller at the right (cavity filled with brine) than at the left (cavity of gas). They can draw some useful conclusions for their applications. However, to know quantitatively the anomalies of the site, one has to solve an *inverse problem* for determining the characteristics of the cavity (depth, geometry and density of the filling). For this purpose, one must have much more micro-gravity data gathered, not only on the ground, but also unusually at many high levels. In principle, according to the *golden rule* of the mathematician Paul Sabatier, from Montpellier University, one must have measurement data as numerous as the unknowns. This necessary condition is not yet sufficient for solving an inverse problem.

Mathematically, a scalar data measured along a plane g(x,y), which is a function of the two coordinates x,y of the balance lying on the flat ground, does allow the recovering of the anomaly $\Delta\rho(x,y,z)$ in a three-dimensional volume. The inverse problem considered is *ill-posed* and does not have a unique solution. In other words, there is no *one-to-one* correspondence (or mapping) between measurements g(x,y) and unknowns $\Delta\rho(x,y,z)$. Another feature of any inverse problem is that it is still ill-posed even if the number of measurements is in adequacy with respect to that of the unknowns. An infinitesimal variation of the measurement (due to uncertainty) may result in a great variation in the numerical results. This is known as the sensitivity of the results with respect to measurements, or in popular speech, the *butterfly* effect of

E. Lorenz, well-known in Fluid Mechanics, because according to the Navier-Stokes equations, a butterfly beating its wings in France (or Brazil) would induce a storm in the Gulf of Mexico (or Texas). This butterfly effect was predicted a long time ago by the French mathematician Jacques Hadamard.

Assume that the number of measurements is roughly sufficient, the mathematical and numerical problem is as follows. Given the variation of gravity $\gamma(\mathbf{x})=g(\mathbf{x})-g_B(\mathbf{x})$ with respect to the Bouguer anomaly reference $g_B(\mathbf{x})$ where $\mathbf{x}$ denotes the point of coordinates $x_{i=1,2,3}$, $\mathbf{x}=(x_1, x_2, x_3)$, find the density difference $e(\mathbf{x})=X(\mathbf{x})-X_0(\mathbf{x})$ where $X=\rho(x)$ is the unknown true density and $X_0=\rho_0(x)$ is the estimated density field in the absence of anomalies. Generally, the density $\rho_0(x)$ of a solid related to the Bouguer anomaly reference is well estimated, from more or less available geological data. The field solution $X(\mathbf{x})$ is related to measurement data $\gamma(\mathbf{x})$ by the Newton gravity equation, with k being the universal constant

$$\gamma(\mathbf{x}) = \mathrm{k}\int_{\Omega} X(\mathbf{y})\frac{\partial}{\partial y_3}\frac{1}{|\mathbf{x}-\mathbf{y}|}dV_y.$$

We see that the inverse problem is difficult at many levels:

a) To know the Bouguer anomaly of the site $g_B(\mathbf{x})$ or the estimated density field $\rho_0(x)$ which is a delicate problem in itself, which can only be solved by specialists in Geology,
b) equivalently, to have the estimated mean density $\rho_0(\mathbf{x})$, which can be also obtained as unknowns of an iterative numerical approximation procedure,
c) to numerically solve the inverse problem.

The first two problems are more or less solved by experienced workers on the site (for measurement and examination), while the third one is essentially a mathematical and very ill-posed problem. Finally, we are facing the problem of determining different fields $g_B(\mathbf{x})$, $\rho_0(\mathbf{x})$ and $X(\mathbf{x})=\rho(\mathbf{x})$ from the data $g(\mathbf{x})$ gathered on the archaeological site. To see well the difficulty of the inverse problem, consider the problem of finding a cavity for which $e(\mathbf{x})=X(\mathbf{x})-X_0(\mathbf{x})<0$ (because $\rho(\mathbf{x})=X(\mathbf{x})=0$ for a cavity and hence $e(\mathbf{x})=-X_0(\mathbf{x})<0$. Mathematicians tell us that, for known $\gamma(\mathbf{x})$, one may have a *ghost* solution in which $e(\mathbf{x})$ is positive, instead of the expected negative value for a cavity, *i.e.*, the presence of a heavier zone than the expected one. It suffices that there are some errors on the data or a lack of numerical accuracy to get ghost solutions. Mathematicians also tell us that one needs to solve the Newton equation with *constraints* $X(\mathbf{x})>0$, by enforcing the numerical solution to satisfy inequality conditions. Otherwise, without physical constraints, the solution of the Newton equation can be meaningless.

There exists a set of numerical tools for solving inverse problems, with specialized names: Tikhonov regularization method, Least Square Method, Weighted Least Square Method (WLS) or Menke method, with or without *a priori* knowledge, Gaussian Stochastic Inversion (which is strictly equivalent to the WLS method for the linear case), Backus and Gilbert method, Simplex method, Karmarkar Method.

It is well beyond the scope of this book to enter here into the details of these methods, which can be found in textbooks and allows us to avoid the butterfly effect. The study of the internal structure of the Cheops pyramid gave to EDF researchers an opportunity to go farther than the simple exploration of the archaeological site. The EDF works on the Cheops pyramid are cited in the paper in French [Bui, 1996] which evoked the joint encounter of Mechanics and Applied Mathematics with Archaeology [H.D. Bui (1996). "La mariage de la mecanique et des materiaux". *La Vie des Sciences,* t. **13**, N°5, pp. 403–407] ["The wedding of Mechanics and Materials"].

The Working Conditions in the Cheops Operation

Under the Technological and Scientific Sponsorship of EDF, created by Marc Albouy of the R&D Division, EDF carried out many works for Archaeology, for example the reconstruction of the stone puzzles, known as the Akhelaton Talatat, performed by Gondran and Vergnieux (1997) and for historical remnants of wrecked boats, the restoration of objects recovered from the boats *Titanic*, *Le Patriote* and *L'Orient* brought out of the sea, Albouy (1994). These boats were sunk near Alexandria (Egypt) during Napoleon's expedition. After these works, in April 1986 EDF received a request from Mr Guillemin, of the Department of Foreign Affairs of France (Cultural, Scientific and Technical Division) for a study on the Cheops pyramid. In essence: *Is it possible to confirm or not the hypotheses of two architects, Gilles Dormion and Jean-Patrice Goidin, on the existence of unknown entrance, corridor and chambers inside the Cheops pyramid, in the North of the King's Chamber*? (unpublished study N° 1).

After eliminating many "hard" methods which present some risk of damage to the pyramid, such as seismic exploration by explosives, drilling or heavier method like the Radar Echo technique (considered by a Japanese team with some success in 1987), EDF proposed the microgravity approach which was well used in its nuclear power plants construction programmes. At the beginning, we were modestly concerning exploration, like the one considered in the Radar Echo technique [(6)].

A mixed experimental team EDF (Jacques Montluçon and Pierre Deletie) and CPGF (Jacques Lakshmanan) were in charge of measurements of gravity in the pyramid site. Jacques Montluçon, *charge de mission*, was precisely in charge of the relation with the Egyptian authority for obtaining necessary authorization for access to the internal part of the pyramid. The mainspring of the first geo-mechanical operation was Pierre Deletie, who was a EDF geologist. He made detailed observations in all accessible parts of the internal pyramid in order to explain the presence of cracks in the monolithic granite beams of the King's Chamber. He was able to tell us that here, we were faced with a soft, porous limestone, containing remnants of empty shells, having such a density or such a hardness or over there, it was the white and fine grain Tourah limestone containing siliceous inclusions, hence an extremely hard stone. He could tell us that the granite of the King's Chamber is a classical porphyry granite stone, slightly denser than the Tourah limestone.

At the same time, another theoretical EDF team, leaded by the author, with his post-doctorate students [(4)] and the theoretical CPGF team leaded by J. Laskhmanan are in charge of identifying the most appropriate numerical strategies.

We tested the stochastic inversion method of Tarantola, with our post-doctorate students and found conditions for it to be the usual simpler Menke method. The latter method was used later in the third Cheops operation (the whole pyramid structure study). The problem was to solve the Newton gravity equation, written in the discretized form $A\mathbf{X}=\mathbf{b}$, « with constraints » on $\mathbf{X}$ (a vector representing the density field at different points) with *a priori* knowledge $\mathbf{X}_0$ which may be *evolutionary* when finer and finer finite elements models were considered (iterative method) and with a weighting factor of 50% for each term of the Menke functional. The pyramid was first considered as one finite element. Then the corresponding solution gave the *mean density* of the whole pyramid, which was considered as the *a priori* knowledge $\mathbf{X}_{01}$ for a refined model of order (n) with n elements, hence the new *a priori* knowledge $\mathbf{X}_{0n}$ of the updated model with (n+1) elements *etc*. This iterative procedure was aimed at solving a nonlinear inverse problem. Alain Bossavit, a scientific advisor of EDF, sent us a Note on the convex analysis of the gravity equation, via the *dual* approach by the «exterior », similar to the dual theory of limit loads in Plasticity introduced by Professor J. Salençon of Ecole Polytechnique [(5)].

Alain Bossavit's Note was entitled « *Si Kheops m'etait conte..* » in the manner of the film « *Si Versailles m'etait conte..* » by the French writer Sacha Guitry. Bossavit's Note indicated that gravity measurements performed along the horizontal corridor to the Queen's Chamber could *never* give the exact geometry of the unknown cavity (object of the second operation on the Cheops pyramid). One is convinced that no adequacy exists between density $\rho(x,y,z)$ and measurement $g(x)$ along a corridor. Moreover, the Newton equation alone is not sufficient for solving the microgravity inverse problem since there are constraints on the density distribution to be considered. The dual approach tells us that the cavity may be located in some region and that we cannot get its exact geometry, using measurements along one axis or along a plane. This is the true reason of the failure of the second Cheops operation on the unknown tomb of the King. If the theoretical team was in Egypt, « *Si Kheops m'etait conte..* », we could have shown Alain Bossavit's note to the experimental team and avoided the media failure of the drillings!

It was the first time that one tried to search for a cavity by microgravity measurements in a historical monument. Our partner, the CPGF, was sceptical about the true possibility of localizing a cavity in this pyramid at least because of the aforementioned reason, $X=0$ in the cavity and $X>0$ outside the cavity. If one expected that there was no stone with density higher than that of the granite, then one might take into account the condition $X \leq 2.65$ T/m3 as well, which might well complicate inversion purposes. (To simplify, we did not consider the last inequality, which might be checked *a posteriori*).

The solution with condition X>0, except in the cavity where X=0, belongs to a class of discontinuous solution which is very difficult to solve if we do not take account of these constraints in the algorithm of resolution.

We just finished the study of the unknown King's Chamber with eventually the treasure to be discovered, under the high pressure of the media. The King's Chamber operation ended by a great disappointment of the negative result. A request of Dr Ahmed Kadry, President of the Egyptian antiquity organisation, to study the whole pyramid was sent to us. But the last operation on the whole pyramid was abandoned by our Directorate after the media failure of the King's Chamber operation.

After a long time, we realized that the failure of the King's Chamber operation (Problem N° 2) was caused by our own work organization. The two EDF teams, the first one in charge of measurements by J. Montlucon and another one for computations by the present author, had no coordination between them. The author had not been associated to measurements. He was not invited to the Paris meeting of October 23, 1986. One reason was that the expedition in Egypt was costly and thus reserved only for administrative CEO who had in mind the exploration operation, but did not have an idea about the aforementioned one-to-one mapping. At least, the author's team could suggest the zones for measurements. Definitively, it is better like that, because measurements were performed objectively, even though imperfectly and independently of the numerical team!

- Measurements had been unnecessarily done in some zones,
- the experimental team ignored the one-to-one mapping between measurements and unknowns, which we will discuss further, by considering simply measurements for the exploration of site and not for the inversion of data,
- there were not enough data in some other zones.

The Blind Test

Let us tell an important anecdote of the Cheops pyramid studies, which was the starting point of fruitful cooperation between EDF and CPGF, even after the failure of the second operation on the unknown the King's Chamber. The geophysical Company considered that the search of a small cavity inside the pyramid, deduction being made for known cavities (corridors, underground chamber, King's Chamber and Queen's Chamber *etc*) resembles the problem of finding *a needle in a haystack*. The unknown cavity might have the expected volume $5\times5\times5=125$ m^3 while the pyramid volume is about 2.5 millions m^3. The number of measurement points was limited to a thousand, because measurements were costly and the number of measurements was fixed by the overall cost of the Cheops project.

Did having a thousand of data measurements imply that the number of finite elements could not be greater than one thousand, which was indeed insufficient for modelling a small cavity? (We should have 50 000 elements at least). Was it possible to use a smaller number of measurements despite the necessity of an one-to-one mapping? To convince CPGF to cooperate with us in the inversion problem, we asked Jacques Lakshmanan, who prepared a PhD thesis in Geomechanics at the University of Nancy, to give us the *theoretical* measurements of gravity (using Nagy's formula) along five (instead of eight) ridges of a homogeneous unit cube, containing a small cavity which was *hidden* to us somewhere in the inside.

This was the *blind test* of March 30, 1987 to show that EDF could recover the cavity despite the insufficient number of measurements (about 50 data for 1 000 finite elements).

A copy of the CPGF listing of calculated gravity along the ridges is given in Fig. 2.4. There were 50 gravity data while the cube was divided by the meshes of $10\times10\times10=1\,000$ finite elements.

By considering the constraints $X\geq0$ and exploiting the mathematical property of the gravity equation, we recovered the centre and the volume of the cavity exactly. On the other hand, the shape of the cavity is not well recovered. If we had been told that the cavity shape was a cube (*a priori* knowledge) then the solution for regular cubic meshes was exactly obtained. Other similar blind tests were positively done, including the test on a two layer material. *A priori* knowledge is a very important data for getting more surely the solution of an inverse problem.

Usually, there is the same number for unknowns and data. Here, owing to the three-dimensional character of the inverse problem, the relative positions of measurement points introduce in some way some complementary and hidden data.

The blind test is apparently characterized by the under-determination of the inverse problem since the data number is less than the number of unknowns. Nevertheless, the blend test was positive and encouraged CPGF to pursue with us on the 3rd problem for studying the whole pyramid. Despite the stopping of the second Cheops operation, the work condition for the third operation was better than for the previous study of the King's Chamber.

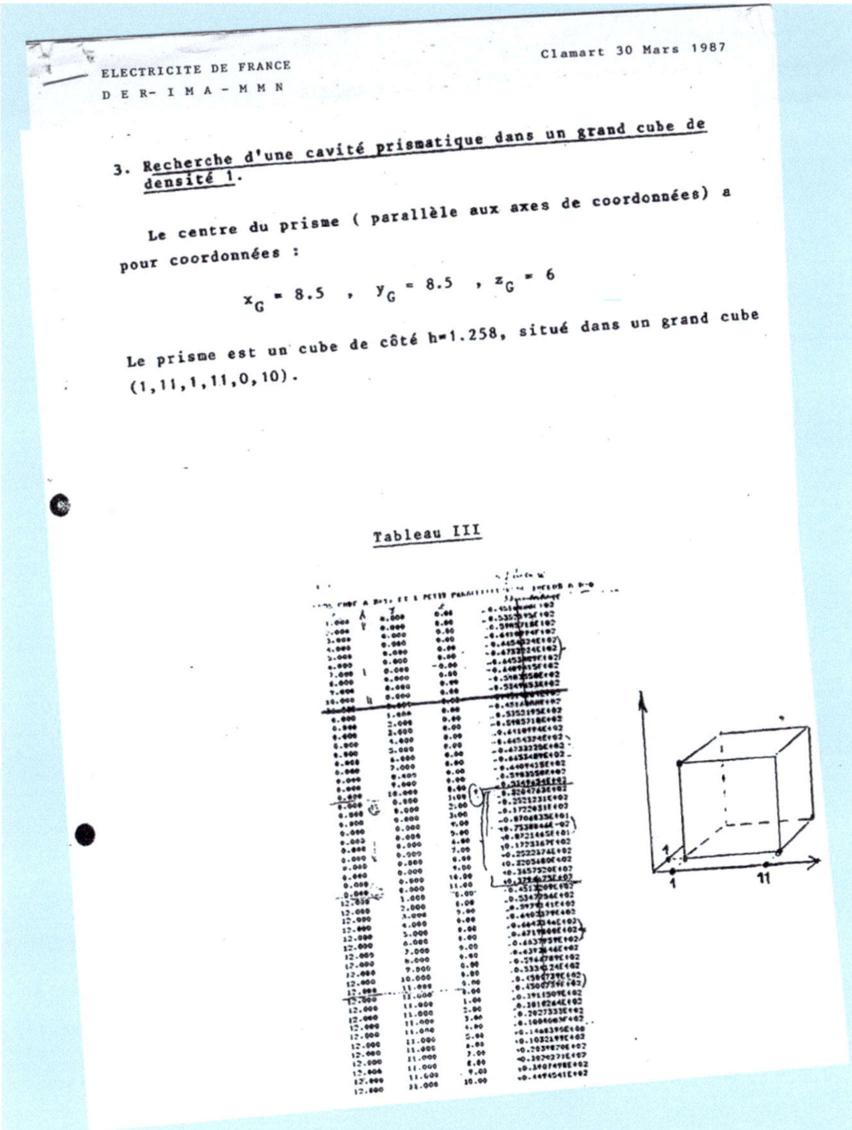

ELECTRICITE DE FRANCE

D E R- I M A - M M N

Clamart 30 Mars 1987

3. **Recherche d'une cavité prismatique dans un grand cube de densité 1.**

Le centre du prisme (parallèle aux axes de coordonnées) a pour coordonnées :

$x_G = 8.5$, $y_G = 8.5$, $z_G = 6$

Le prisme est un cube de côté h=1.258, situé dans un grand cube (1,11,1,11,0,10).

Tableau III

[illegible]

Fig. 2.4 Copy of the numerical blind test result of March 30, 1987 to find the unknown cavity from theoretical data provided by CPGF, along 5 sides of a cube of size 10. The centre of the small hidden cube (x_G, y_G, z_G) of size 1.258 is exactly recovered by the inversion of data

Chapter 3
Density Images by Microgravity

My work always tried to unite the true and the beautiful, but when I have to choose one or the other, I usually choose the beauty.
H. Weyl.

Microgravity is a technique that initially addressed the exploration of sites which does not have the many restrictive conditions of the electric exploration method. Jacques Lakshmanan (1963), with Jean-Claude Erling, was the first to introduce the micro-gravity technique in France, for many applications: study of foundations, cavity detection in the SNCF railway embankments and heavy deposit ore in underground. EDF approached him for measurements on the sites of its future dams. In the 50's the microgravity measurement did not exist yet. With such a method, one could have detected the fluid flow under the supporting foundation of the Malpasset dam (in the Southern France) and prevented its catastrophic failure in December 1959.

In December 1985, architects Gilles Dormion and Jean-Patrice Goidin made a detailed study of the anomalies in the construction of the Cheops pyramid, on the walls of the King's Chamber and in the horizontal corridor to the Queen's chamber, which might indicate possible presence of unknown corridors and chambers in the Northern side of the King's Chamber. With the support of the Department of Foreign Affairs of France, they made complementary works on the Cheops site and found an unknown system of doors. They asked for a microgravity exploration in support of their discovery. Jacques Montlucon was contacted by the French Embassy at Cairo and then rapidly EDF, using its Technological and Scientific Sponsorship and its usual partner for gravity measurements, the CPGF company, decided the first operations on the Cheops pyramid, with the approval of the Egyptian Antiquity Committee at Cairo. The first operation was on the mechanical analyses of cracks appearing in the five monolithic granite beams of the King's Chamber and the second one on the unknown King's Chamber using microgravity measurements.

The CPGF Company was an expert in microgravity measurements while EDF possessed the know-how of its engineers and researchers for mathematical and numerical analyses, its computing facilities (the Cray 1 computer) and above all

H.D. Bui, *Imaging the Cheops Pyramid*, Solid Mechanics and Its Applications 182,
DOI 10.1007/978-94-007-2657-4_3,

the financial support. The Cheops pyramid project was spread on two years 1986-1987 and consisted of many successive steps corresponding to four measurement campaigns on the Cheops site, with the total number 754 measurement points along internal corridors, chambers, ridges and bases of the pyramid. The Lacoste and Romberg *microgals* balance was used for measurements. The reason for the micro-gravity measurement method is manifold; efficiency in the cavity research, soft non destructive method of detection, possibility of data inversion. The Radar Echo method is also a non destructive detection method for the exploration. Its use for data inversion is rather delicate (6).

It was the first time that a true imaging was done on the Cheops pyramid site. An imaging implies real measurements and data inversion of the gravity equation. The difference between an imaging and an exploration of sites is as large as between a medical scanner (7) and an ordinary radiography (the scanner implies measurements in many directions of X-rays and inversion, while the radiography involves a measurement in one direction of X-rays without inversion).

The Second Solar Boat Discovery

The efficiency of the Lacoste and Romberg *microgals* balance was demonstrated by a test near the museum of the first Solar boat (1945). An anomaly of −30 to −50 μGals, in the South of the Cheops pyramid confirmed the presence of the cavity containing the second Solar boat discovered previously by Archaeologists. An oral presentation of our work on the second Solar boat was given in the Symposium on Egyptian Antiquity at Cairo [Bui *et al*, 1987]. The Solar boat is still in its grave for fear of damage was it to be brought out to the open air. Shortly after the discovery, an American geographic team made a film on the boat and since 1992, a Japanese archaeological team studied the means for its protection from humidity and Sun light. Recently, Japanese researchers introduced a mini camera for a public show of the Solar boat. After the success of the second boat localization, the Egyptian authorities encouraged EDF and CPGF to pursue the Cheops project farther. This was new in the sense that for the first time an attempt was made to measure, not only the variation of gravity which gave immediately qualitative results for an exploration, as for the case of the second Solar boat, but the density field in a structure. More exactly, it concerned *the inversion of gravity data* in order to obtain a density *imaging* of the pyramid. Could we find this way a cavity or an unknown chamber in the Cheops pyramid ?

Chronologically, there were two microgravity studies on the Cheops pyramid: research of the unknown King's Chamber and our study of the whole pyramid. These studies were different in time, duration and issues at stakes. The second one required new mathematical and numerical methods and needed much more time. The first study was done in a hurry while the media looked on, waiting for the revealing of the unknown tomb and the treasure of the King and the second one was performed with discretion in the manner more consistent with the spirit of scientific

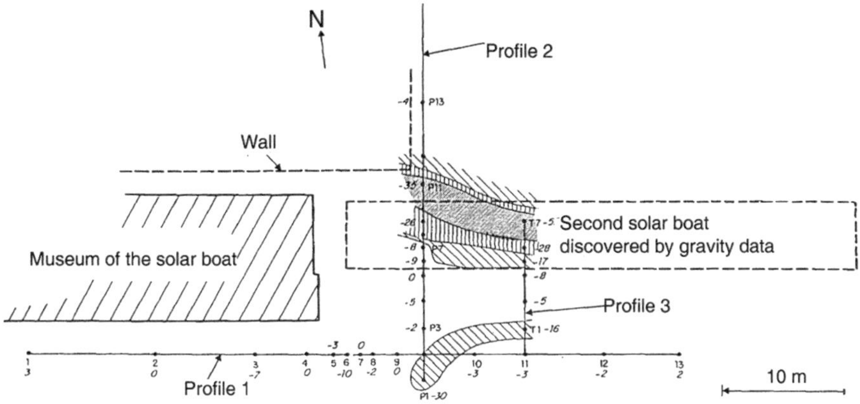

Fig. 3.1 Exploration of the second Solar boat next to the first Solar boat Museum

research works, done in serenity for a publication in scientific journals rather than a communication to News agencies.

The Measurement Campaign in the Pyramid Site

The measurement campaign for the King's Chamber operation gained from the aid of the French Embassy at Cairo, who asked the *Societe Generale d'Entreprise*, the French builder of the underground metro of Cairo, to provide us some scaffolding in the Great Gallery and the King's Chamber. Previous measurements for the unknown King's Chamber were done along the horizontal corridor to the Queen's Chamber and did not need scaffoldings. Measurements were done on the North and South walls of the Great Gallery, on the inclined floor, the inclined corridor and in Cavaglia's hole. Captain Giovanni Battista Caviglia (circa 1817 AD) used explosives to blast holes in the Cheops pyramid, searching for chambers, unsuccessfully. Other measurements were made along the horizontal corridor at two height levels. Measurements in the King's Chamber were done for 3 height levels and 3 rows and in the five « discharge chambers » for one height level and three rows. Denser measurements in the King's Chamber were aimed at the direct comparison with a calculation of gravity with and without an assumed cavity. A synthetic view of 218 measurements stations near the King's Chamber is shown in Fig. 3.2.

Measurement Results Near the King's Chamber Structure

Figure 3.3 indicates the results of gravity anomaly in the King's Chamber, *i.e.* the gravity $g(x_1, x_2, x_3)$ indicated by the balance which depends on the altitude x_3 deducted for the influence of the pyramid P, the presence of known voids V and the regional correction of the Giza plateau (which constitutes the Bouguer correction g_B). The Bouguer correction was the gravity existing "before" the pyramid building and was of the form $g_B(\mathbf{x}) = ax_1+bx_2+c$ at the ground level $x_3=0$, which varied a little with the coordinate x_3 in comparison with the larger influence of the future construction. In contrast with the case of an underground cavity, where the density $\rho_0(\mathbf{x})$ is not yet known, the initial density existing before the construction at points $x_3>0$ vanishes $\rho_0(\mathbf{x}) = 0$.

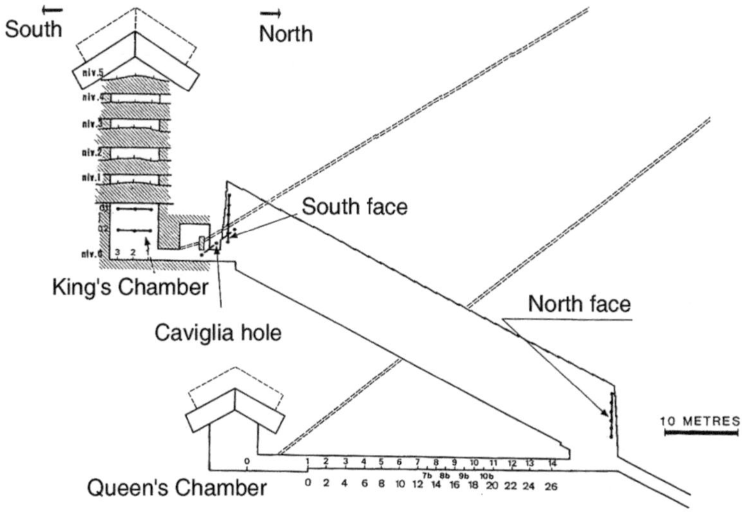

Fig. 3.2 South, North, King's Chamber, South face Profiles, measurements in Cavaglia's hole, Queen's Chamber, North faces profiles, EDF-CPGF

After the construction, the influence $F(x, P, \rho)$ of the pyramid from the density distribution $\rho(\mathbf{x})$ inside the pyramid P at each point $\mathbf{x}$ is defined by the integral $F(x, P, \rho) = k \int_P \rho(y) \frac{\partial}{\partial y_3} \frac{1}{|x-y|} dV_y$ which depends on the geometry of P and the density $\rho(\mathbf{x})$. Inside the pyramid we have the relation $g = g_B + F(\mathbf{x}, P, \rho)$ which allows the determination of the density distribution $\rho(\mathbf{x})$ by solving an inverse problem with data $g(\mathbf{x})$.

But, without any inversion of data, we can get an estimation of the *mean density* very simply, by remarking that the volume of the unknown cavity C do not excess the value 200 m^3 while the known cavities V (corridors, chambers, gallery) represent the total value of V=2 000 m^3, both C and V can be neglected in comparison with the pyramid volume of about 2 340 000 m^3 (by adding 1 m thick missing finishing stones, one has 40 000 m^3 more). We can legitimately neglect the influence of C and V ($\rho=0$ there) in the integral $F(\mathbf{x}, P, \rho)$ and apply the « mean value theorem » for calculating the integral $F(\mathbf{x},P,\rho)$ as follows

$$F(x, P, \rho) = k \int_P \rho(y) \frac{\partial}{\partial y_3} \frac{1}{|x-y|} dV_y = k\rho_m \int_P \frac{\partial}{\partial y_3} \frac{1}{|x-y|} dV_y$$

where ρ_m or d_m is the mean density of the whole pyramid. Let us denote the integral $k \int_P \partial |x - y| / \partial y_3 dV_y$ by $F(\mathbf{x},P)$. We obtain the formula

$$g(\mathbf{x}) = g_B + F(\mathbf{x},P)d$$

which establishes a simple relation between the influence P of the pyramid and $d=d_m$. A more precise formula can be obtained with the deduction of known cavities influence (King and Queen's Chambers, Great Gallery, corridors). But at this stage, it is not necessary to do so.

The Low Mean Density 2.05 T/m^3 of the Pyramid

In the plane of abscissa F and coordinate g, we plot different gravity measurements $g(\mathbf{x})$ at different points $\mathbf{x}$, as functions of the influence $F(\mathbf{x}, P)$, which can be easily calculated from the *geometry* of the pyramid. We see that according to the equation $g(\mathbf{x}) = g_B + F(\mathbf{x},P)d$, different points (g, F) may be aligned along a straight line the slope of which is exactly the *mean density* d. Real measurements agree well with the theory, Fig. 3.4. Using this simple method, with measurements at different levels (descending corridor, ascending corridor, King's Chamber, discharge chambers) we obtained the mean density $d_m=2.05$ T/m^3 very accurately, *i.e.* a density much more low than that of ungraded limestone (density 2.35). This surprising result obtained from measurements, without any inversion of the equation (more exactly with a division between two numbers) but only with direct computations of integrals $F(\mathbf{x}, P)$ demonstrated that there were so many voids in the pyramid.

One can observe behind missing stones at the bases and corners of the pyramid that there are many voids between stones and that visible stones are roughly squared, far from the perfect parallelepiped. Therefore there are many voids which explain the low overall density. We shall confirm this important point by considering later a very simple experiment which anyone can do.

One can make comparable observations on the Cheops pyramid and on other pyramids. In the wall of the Meidoum degrees pyramid, consisting of a masonry or parallel stonework joined by mortars, there is a huge hole which unveils disordered and non squared stones, more or less joined by plastering. One wonders how these stones were in their place if they were not joined by mortars. Perhaps, disordered stones can be blocked mutually because of their complex 3D geometry, just like the puzzle of stones on the top platform which are more rigid than a well-ordered and squared and faced array of stones or the puzzle of stones of some walls, for example walls in the Low Khephren temple better resist to an earthquake than horizontal layers of squared stones.

Fig. 3.3 The measurements of anomalies of gravity in the King's Chamber (deficits of gravity in blue and excess in Red). The deficit of gravity does not take any particular form suggesting a cavity or a heavy mass or not. Only an inversion of the results of measurement can give an indication of the true deficit of density (EDF-CPGF)

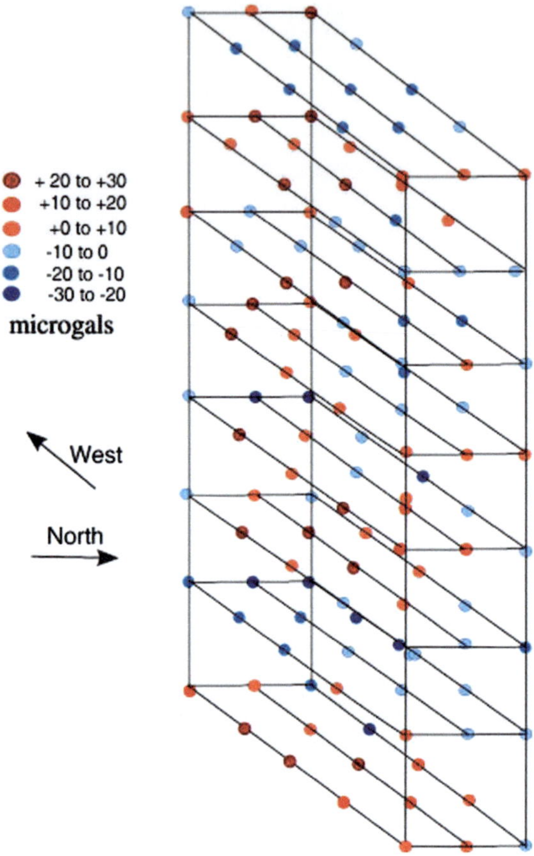

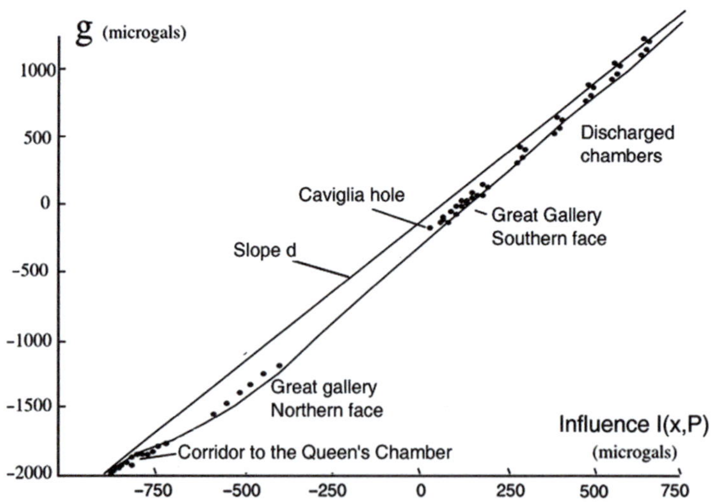

Fig. 3.4 Direct evaluation of the average density, by the slope d (EDF-CPGF, 1987)

Fig. 3.5 The Meidoum pyramid. The degree, revealed by the slip of added blocks to this face of 32 m height, shows a huge hole and behind it, the thickness of the masonry wall and filling unsquared disordered stones with much of vacuum (Permission Ch. Hachet)

Interstices and Voids

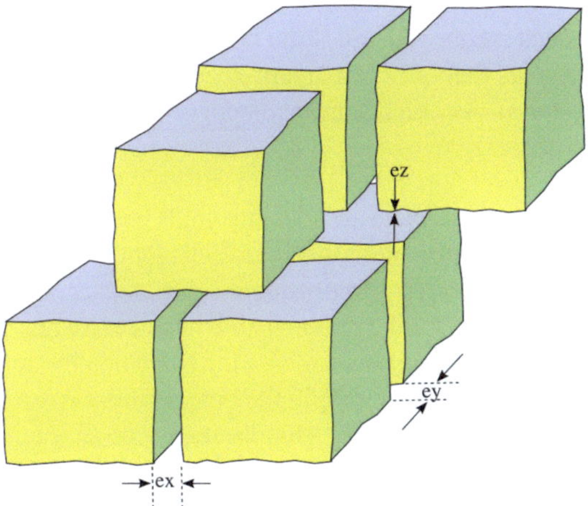

Fig. 3.6 Interstices in 3 directions. For perfect vertical joint and equal horizontal interstices, the density is given below

An arrangement of cubes of equal height with the interstices e_x, e_y, e_z in 3 directions, has the apparent density:

$$d_{app} = (1 - e_x)\left(1 - e_y\right)(1 - e_z)d,$$

(we use here x, y, z instead of x_1, x_2, x_3). In the case of stones perfectly squared in their horizontal face, e_z=0, in order to better transmit the vertical pressure and for e_x=e_y=e and d=2.35 we have the density given in the table.

e(%)	void(%)	d_{app}
1	2	2.30
5	10	2.12
10	19	1.90

Fig. 3.7 Box of 3 layers of well arranged sugar cubes. Two layers are sufficient to fill the box with disordered cubes

A full box of cube sugars contains originally three layers of sugars. Once the cubes are moved out of the box, we put them again in the box without any order. It appears that only two layers of cubes suffice to fill the box, *i.e.* there is 33% of void (Photo). A compact arrangement of spheres [(8)] with the same diameter has a weaker percentage of void $1 - \pi/\sqrt{18} \approx 26\%$. The percentage is smaller when there are different diameters.

Therefore there are so many voids between stones, which can be created either by the dissolution of plasters between stones, or simply by existing voids due to irregular shape stones, like disordered cube sugars in our box of Fig. 3.7. In 1819, G. A. F. Fitzclarence broke into a visible huge cavity near the North-Eastern ridge, about 87 m height, which was called « the fox hole» by our colleague Pierre Deletie when he saw a fox disappearing in the hole. Is this cavity, which was recently revisited by B. Brier, the notch of the Houdin theory of an internal ramp? There is another theory which involves such a notch or a cavity, for example the platforms of type A of the Holscher zigzag ramp theory badly filled up with stones, which will be discussed in Chapter 4. The photos shown in Houdin (2009) indicate masonry stones nearby piles of blocks with huge voids between them. At the beginning of the Cheops operations, Pierre Deletie, geologist at EDF, told us that there were too many voids. The first part of the construction seemed to be done carefully. For example, the fault across the limestone foundation was accurately filled in with limestone; the finishing squared blocks of stones at the bottom were laid down without interstices, *etc.* At the level of the King's Chamber, the visible masonry around the King's Chamber and the discharge chambers showed squared stones. One can see voids larger than one hand and sometimes there is a draught, as noted by J. Kerisel (1991), which means that air flow passes through connecting interstices between stones.

It is plausible that voids resulted from the dissolution of plasters by the rain, as noted by Kerisel (1991) who quoted an ancient author « *The water from the sky is the wrath of the God Seth, it is harmful* ». This phrase indicates the devastating flood by the rain in Egypt at this time. In the past, the rain was more abundant than today and morning dewdrops were daily. We quote again Kerisel who cited a text of Pline « *the sky spring is the saliva of the stars* ».

A void of v=10% corresponds to the density d=2.12 T/m^3 and the void v=19% corresponds to the density d = 1.90. Even if the stones are perfectly squared on horizontal faces, ensuring perfect vertical contact between stones and hence a better stability of the structure, a vertical interstice of e=10% results in the density d=1.90 T/m^3 (lower than the density of plaster d=2). If the Egyptians wished to build a light and solid construction, they had those efficient methods to do so: to space out stones laterally and to have perfectly squared stones horizontally.

In examples of voids between blocks, we see that void depends on the shape and the size of stone blocks. The void can be very important when stones have the same shape and size and if the stone arrangement is not periodic.

Conversely, to minimize the void volume, one considers the arrangement of spheres of any size smaller than the voids existing between larger spheres. So that in any construction, including in the Cheops pyramid, one fills in small materials and plasters between stones. One can find in the Cheops pyramid, the filling in with small stones, bricks, sand and plasters. The dissolution of plasters explains the low mean density of the pyramid as revealed by our microgravity measurements. The Meidoum pyramid which shows a huge hole on its wall of height greater than 32 m, Fig. 3.5, was a modified pyramid by addition of stones on the faces of the former pyramid. Such a structure had very low cohesion with the ancient inner core, because of missing links between old and new structures. Without any doubt, the added part of the pyramid was unstable and collapsed by an instability mechanism, in the sense of Soil Mechanics, see J. Salençon (1977) [(9)], maybe induced by an earthquake as suggested by some authors.

Consider the outer layer of the pyramid built by putting squared stones layer by layer, except on the inscribed zigzag ramps of the Holscher or Goyon theory. The ramps themselves were built with compact masonry of joined stones. But the filling materials above the ramps could be spaced stones with many voids. Consequently, the apparent surface density is lower around or above the ramps. Probably, the theft of stones or bad filling up of stones in these parts may be the cause of the collapse of some part of the pyramid, as observed for example near the S-E corner.

Direct Computation of Gravity due to a Cavity

Result of Fig. 3.3 does not allow us to guess what is hidden behind gravity measurements which are apparently not linked to the presence of cavities.

The excess of gravity can be explained by the granite monolithic beams the length of which may not be well estimated. It is then interesting to look at the *direct*

Fig. 3.8 Visible missing stones at the S-E corner. (Permission of Tatiana, http://tatiana.blogs.com)

influence of a hypothetical cavity located near the North side of the King's Chamber, as expected by some architects. Theoretical values of gravity due to the cavity (in Green) are shown in Fig. 3.9. As expected, one obtains a continuous variation from negative value −30 μGals in Blue, to positive one +30 μGals in Red (there is a lesser attractive force because of a missing mass above the apparatus). Comparing the direct computation, Fig. 3.9, with measurements, Fig. 3.3, we see some differences, but we are unable to draw a conclusion on the existence of cavities or not. A direct computation by trials and errors of the geometry and location of a cavity has no chance to get the solution of an inverse problem. Did a cavity with a thick granite wall could be detected? The high density of a granite wall is compensated by the cavity so that the anomaly of gravity would be not significant. What is the criterion for a « solution » to be a good one or not ? Again, an exploration does not give us precise information on the unknown object [(10)]. It looks like the situation of our doctor who palpates our body without drawing a conclusion and then sends us to the hospital for complementary examinations by an X-ray Scanner or a MRI (Magnetic Resonance Imaging). Methods of Medical Imaging made use of the data inversion by computers. Like the exploration in Geophysics, the examination by our doctor is based only on his own experiences. Like methods of imaging in Medicine, we shall make an imaging of the pyramid by considering the inversion of the Newton equation with gravity data.

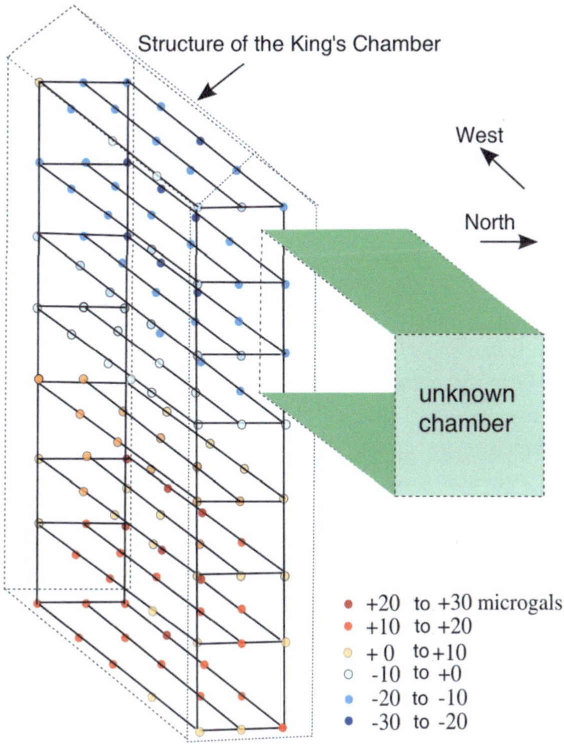

Fig. 3.9 Cavity in Green placed beside the King's Chamber, (EDF-CPGF)

Inversion of Gravity Data for Cavities Near the Chambers

We see that the comparison between a direct computation of the influence by a hypothetic cavity and experimental data is not conclusive because of the possible presence of other internal anomalies which should have some importance (cavity with thick granite wall). For example, we might overestimate the granite monolithic length and thus obtain numerically the ghost deficit of density somewhere, by compensation between different terms of the mathematical functional discussed below. We decide to solve an inverse problem after testing different methods of iterative solution.

We have a precious information about the mean density of the whole pyramid d=2.03 to 2.05. It is the *a priori* knowledge introduced in the preceding Chapter. Without *a priori* knowledge, it is difficult and even impossible to solve the inverse problem.

Let us take a simple example of common life which is the *Rally game* at TV shows We are asked to find the Rally message at the lowest point of the Bievre Valley at the South of Paris. We follow the Bievre River up to the lake near the Heller park at Antony, which is a control basin where the river becomes an underground tunnel connecting to the Seine River at Paris. An underground tunnel is no longer a river. The message has to be found in the Heller lake area. In the same manner to seek the lowest point of the Bievre river which is the Heller park lake, we search the "solution" of the inverse problem defined as the lowest point of a mathematical "functional" defined hereafter, which takes account of the "constraints" or inequalities on densities. This mathematical process is called an "optimization problem with constraints".

Schematically, when we divide the pyramid P into small blocks or finite elements, numbered as n=1, 2, ... N, with the total number N, we can set a mean elementary and constant unknown density d_n or $\rho_n \geq 0$ to each element, the ensemble of elementary elements ρn defines a vector $\mathbf{X} = (\rho_1, \rho_2 \ldots, \rho_N)$ called "unknowns" of an Euclidian space R^N of N dimensions, with components $X_n = \rho_n$.

Gravity measurements at M points define another vector $\mathbf{G} = (G_1, G_2, \ldots, G_M)$ of the Euclidian space R^M. The difference between measurement **G** and the Bouguer correction g_B constitutes the « data » vector called **b**. Finally, the Newton gravity equation $g(\mathbf{x}) - g_B(\mathbf{x}) = F(\mathbf{x}, P, \rho)$ yields a linear system of equation of the form A.**X**=**b**, where A represents the rectangular influence matrix, of M rows and N columns, which results from the discretization of F(**x**, P, ρ), *i.e.* an operation resulting from the decomposition of the pyramid into finite elements. This *linear* system of equations for the unknowns **X** is augmented by the « constraints » on the unknowns, for example $\rho \geq 0$ and $\rho \leq \rho_{\text{granite}}$ and therefore becomes a *nonlinear* system. Moreover, the mean density is about d=2.05 T/m^3. The overall nonlinear system of equations and inequalities gives rise to a convex analysis problem, for which one has many mathematical methods. That is to minimize some functional in a *convex* ensemble. Let us indicate some elementary Mathematics of the gravity inversion without going into the details which are beyond the scope of the book.

Some Mathematics of the Inversion

The optimal solution minimizes a functional related to the magnitude of the *residuals*, for different methods, which measure the difference between the model of data AX and the observation one **b**, taking account of the fact the model **X** is near from the knowledge $\mathbf{X}_0$. One has what is called "*an optimization problem under constraints* $\mathbf{X}_n \geq 0$, *etc.*"

$$\underset{X_n \geq 0,\ |X_n - X_{n0}| \leq C}{Min} \left\{ \|AX - \mathbf{b}\|^2 \right\} \quad \text{(Standard method)}$$

$$\underset{X_n \geq 0}{Min} \left\{ \|AX - \mathbf{b}\|^2 + \alpha\, |\mathbf{X} - \mathbf{X}_0|^2 \right\} \text{(Tikhonov method, } \alpha > 0\text{)}$$

$$\underset{X_n \geq 0}{Min} \left\{ \frac{1}{2}\,(AX - \mathbf{b})C_b^{-1}\,(AX - \mathbf{b}) + \frac{1}{2}\,(\mathbf{X} - \mathbf{X}_0)C_X^{-1}\,(\mathbf{X} - \mathbf{X}_0) \right\} \text{(WLS method)}$$

where $\mathbf{X}_0$= (d, d, d,. . .d) is the *a priori* knowledge, C is some constant for bounding the space of solution, $\alpha \geq 0$ is the Tikhonov regularization constant, chosen here arbitrarily, but possibly determined in an optimal manner by Kitagawa's choice method, (See the reference [Bui, 1993]) and C_X et C_b are covariant matrices for weighting the two terms, in general diagonal matrices and positive matrices of rank N and M respectively, introduced for having some range of the solution **X** and some uncertainty on the measurement **b**. When one does not have an idea on the solution, or on its mean value, the covariant matrix CX which is the *expectation* $C_X = E\left\{\mathbf{XX}^t\right\}$ is large and therefore C_X^{-1} is very small. It becomes equivalent to ignore the *a priori* knowledge in the WLS method (Menke's method). When one does not ignore this term, one may consider Tarantola's method or Menke's method. In practice, we consider Menke's method with simply diagonal covariant matrices.

That is to consider simply the Tikhonov regularization procedure, with positive constraints $X_n \geq 0$ on the unknowns, since gravity measurements by EDF-CPGF teams were performed with high accuracy. Moreover, the calculation of the influence integrals for a parallelepiped can be done analytically with Nagy's formula, see the reference [Bui, 1993].

It is easy to understand the Tikhonov method. Without the regularization term, the positive eigen-values of the matrix are very small, which render the inversion impossible. By adding a positive term $\alpha > 0$ to the functional, the whole eigen-spectrum moves to the positive value and enables the inversion. But if the regularization term is too high, one may loss the physical meaning of the numerical solution. Finally, we quote the mathematician Pierre Sabatier « *solving an inverse problem is an art*», precisely the art of making the best choice of the compromise.

There are many manners to solve a gravity inverse problem, independently of the choice of the mathematical functional. For example, we can consider the unknown density $\rho(\mathbf{x})$ for the whole pyramid. One can also take into account of known cavities, for which $\rho(\mathbf{x}) = 0$ (King's and Queen's Chambers, corridors, Great Gallery, *etc.*). We have seen that the Bouguer anomaly for the homogeneous pyramid P can be defined as $g_P(\mathbf{x}) = F(\mathbf{x}, P, \rho_0)$ with the mean density ρ_0=2.05 and the residual anomaly $C = g - g_B - g_P$ may be explained by the density difference $e = \rho - \rho_0$ or $e = X_n - X_0$ to be determined. For the interpretation of numerical results, it is a talking picture that points e<0 correspond to a *deficit* of density (light zones in Blue or Green) and points e>0 correspond to an *excess* of density (heavy zones in red colour).

Since, one is searching a cavity near the King's Chamber in certain zone Z, we set e=0 outside this zone. The choice of Z is rather arbitrary, but anyway it can be modified by further corrections, in an *iterative* procedure to solve a *nonlinear* problem. For granite beams or walls we set e=2.5−2.05=+0.45, for inner cavity e=−2.05. For other elements, e($\mathbf{x}$) is the unknown as well as the constants a, b, c of the Bouguer anomaly of the site before the construction of the Cheops pyramid to be determined.

The central zone is divided into three vertical blocks, a block containing the Great Gallery and two blocks at its side. The total number of elements in the blocks is 46. In the first computation, we did not consider the inequality constraints. It appeared that the numerical result was stable enough, with the *a priori* knowledge $\mathbf{X}_0 = 2.05$, without violating the constraints. The calculations obtained for measurements in a plane (or two neighbouring planes), are shown Fig. 3.10. There are:

- light zones (A) ($e < 0$) at the centre and above the Great Gallery,
- light zones under the Queen's Chamber (A) and midway below the corridor to the Queen's Chamber,
- a heavy zone (B) ($e > 0$) near and under the portcullis chamber,
- a heavy zone (B) midway above the corridor to the Queen's Chamber,
- a possible light zone (C) midway between the King and the Queen Chambers.

Do light zones A at the Great Gallery level and heavy zones B correspond to cavities and storage (or anti-chambers) rooms in the Western side of the Great Gallery, with thick walls, expected by Architects Dormion and Goidin (1986) **(11)** after noticing some strange details at the King Chamber's walls? The light zone (A) under the horizontal corridor to the Queen's Chamber at the East corresponds also to a cavity expected by the architects. Three inclined drillings (red lines, downwards in the East direction) were undertaken by CPGF at this place of the horizontal corridor for obtaining only some sand of homogeneous grain size instead of a cavity. The origin of the sand is still a matter of controversy. Some trace of salt was found there, a fact meaning the dissolution of plasters by water. Egyptologist Lauer thought about the sand used for filling in stone interstices. Later, a 2D inverse problem solution will not confirm the cavity, but only a zone of light density.

The experimental team involved in the Cheops's Chamber operation erroneously confused exploration and inversion. The confusion in methods to recover a cavity and above all the precipitation explained the failure of the operation on the unknown King's Chamber and the end of the Cheops pyramid operation. Despite the media fiasco of the drillings, retain however an important conclusion. According to Marc Albouy's preface "*After three meters of drilling through a thick wall cladding... and then into a second wall... etc.*" we knew that there were many masonry walls inside the pyramid (possibly masonry walls of degrees, open air tombs, corridors, reinforcement walls like those observed in the Senostris II pyramid *etc.*) A long time after the King's Chamber operation, we thought about the true reason of the failure and the media fiasco of the second Cheops operation. After the success of the exploration of the site containing the second Solar boat, why not to expect a similar success, even a greater one, of the exploration of the horizontal corridor? The true reasons indeed were the precipitation of the experimental team and the confusion between exploration and inversion. In French we said "*Vendre la peau de l'ours avant de l'avoir tué*". [Sell your chickens before they hatch].

Imaging the Pyramid with Microgravity Measurement

Mathematically, the failure of the Cheops Chamber operation can be explained by the following arguments:

a) Measurements done outside the pyramid, along the four ridges and the basis, are too far from the hypothetic cavity to be used as data for the inversion in the central part of the pyramid.
b) Measurement points for the unknown King's Chamber and nearby are roughly in the plane joining the top of the pyramid and the Great Gallery, even if there are measurements in two neighbouring adjacent planes, with the short distance 1 m. The measured gravity is a function of two variables g(x, y) and worse, a function of one variable g(x) along the horizontal corridor. In the first case, measurements in a plane can be used as data for recovering a long cavity or a corridor, perpendicular to the plane, not to search an isolated chamber. These measurements are good for a 2D imaging and not for recovering a 3D object. Their interpretation for the real case is delicate, although there is some correlation between 2D and 3D models.

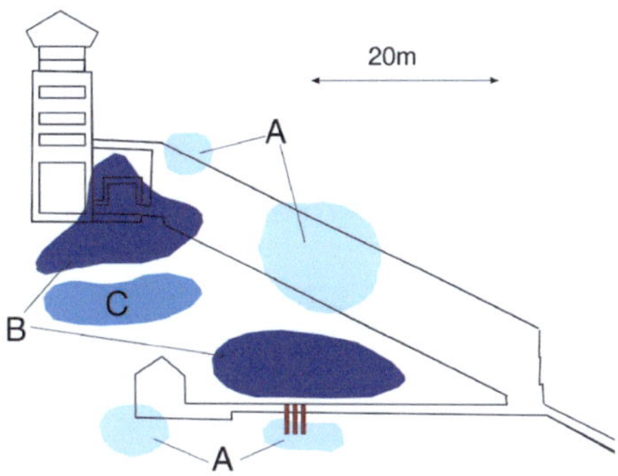

Fig. 3.10 Anomalies of density in the central core. A 2D inversion is made with gravity data nearly in a vertical plane. The 2D results *cannot* be used to give the solution for the real 3D case, although there are some correlations between both theories. The light zone A under the drillings (in Red) is a "ghost solution". A cavity could very well be right next

Measurement data are not adequate enough for recovering the unknown 3D density. It is like the radiography imaging using one parallel X-ray beam. Only with the invention of scanners, which make use of X-ray beams in all directions of a plane, that one has a more precise sliced image of the body.

In contradistinction, the study of the whole pyramid requires all measurements inside and outside it. For example, measurement points on the North-Eastern and South-Eastern ridges and on the Eastern basis are shown in Fig. 3.11. (The image was on the 1987 Season Card from our department MNM/R&D/EDF).

From gravity measurements, it is easy to calculate the residual anomaly $\gamma(\mathbf{x}) = g(\mathbf{x}) - g_B(\mathbf{x})$ along the ridges and the perimeter of the pyramid, by subtracting the gravity $g_B(\mathbf{x})$ corresponding to the mean density $\rho_0(x) = 2.05\,\mathrm{T/m^3}$, *i.e.* the gravity inside the homogenous pyramid P. At this stage, there is no data inversion yet. The residual anomalies considered for the inversion are surprisingly regular along the ridges and perimeter. For example, the residual anomaly is positive (in Red) almost along the Eastern perimeter, which means that there are many voids above the balance. A zero residual should mean an exact solution. There are lesser masses in the pyramid which would attract the mass m of the spring upwards so that the vertical gravity is higher than expected there. Moreover, the top part of the pyramid may have a density much lower than $2.05\,\mathrm{T/m^3}$ to produce a residual gravity so negative (green colour in Fig. 3.12).

The latter point will be confirmed by an inverse analysis. We are now sure that the pyramid is *not homogeneous*, which we have expected only by observing locally the density at some places. The aim of an inverse analysis is to find a *heterogeneous* solid or the density $\rho(\mathbf{x})$ in such a way that the absolute value of the residual anomalies decreases significantly everywhere.

The exact solution corresponds to the zero residual. The quality of the gravity inversion can be observed by comparing the 3D graphics of Fig. 3.13 with Fig. 3.12. It is obvious that the solution for the heterogeneous solid is better than that of the homogeneous one, because the residual, which measures the error on the solution, is much lower in the heterogeneous solution, Fig. 3.13. In the last Figure, some large residual, positive or negative, are observed at some isolated points.

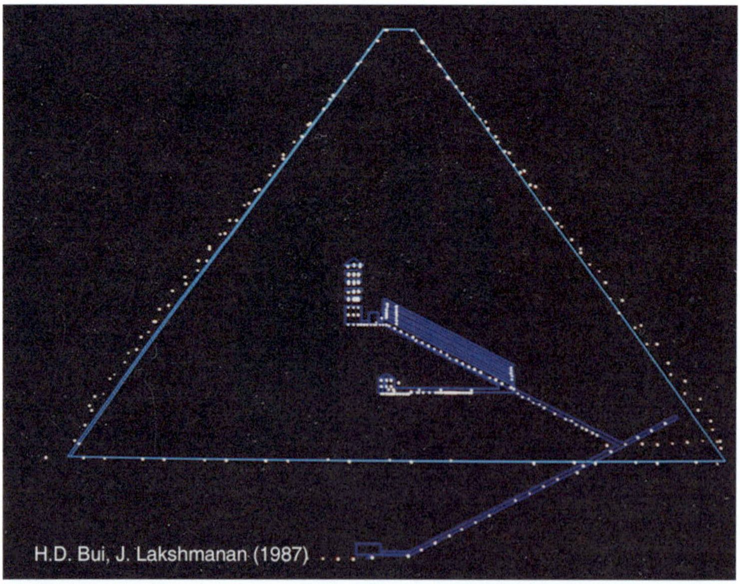

Fig. 3.11 Greetings card 1987 of the Department MNM/EDF-CPGF

Three-Dimensional Meshes of the Pyramid

The pyramid and its foundation are divided into 34 *macro-elements* which are:

- 1 exterior block from the foot of the pyramid to the distance 5 000 m,
- 6 underground blocks,
- 1 big central element,
- 2 blocks inside the central element for modelling the King's Chamber structure,
- The remaining being divided into 24 macro-elements, as shown in Fig. 3.14.

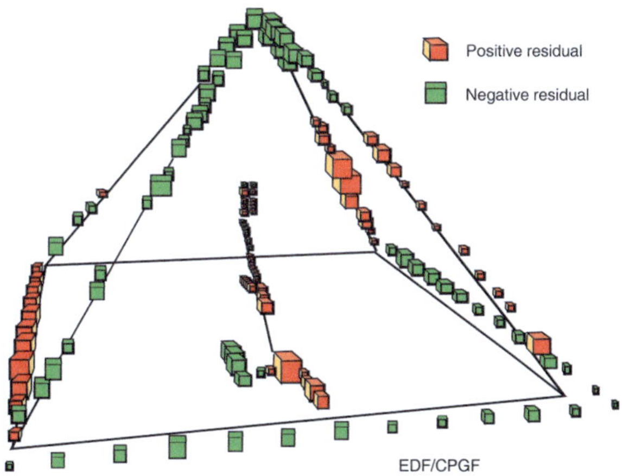

Fig. 3.12 The residual anomaly C=g-gP for a homogeneous model is very negative at the top, what shows that the solution is wrong. There are important deficits of density in top (EDF-CPGF)

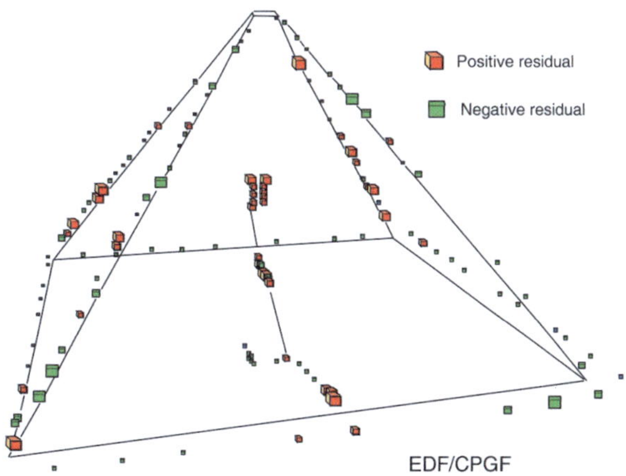

Fig. 3.13 Residual for a heterogeneous model which approaches the true solid. The numerical solution is much better (EDF-CPGF)

In turn, the 34 macro-elements are divided in all into 2 000 *micro-elements*. We then have a *convex analysis* with a rectangular matrix A (754 × 2 000). Remark that the rectangular matrix is not invertible and that the number of data (754) is much lower than the number of unknowns (2 000). As for the blind test, with the matrix (50 × 1 000), we need a special method of inversion.

The computation was done on the Cray computer of EDF at the Clamart site in the Hauts-de-Seine Department and the numerical results were analysed by hands, not that EDF did not have adequate graphical analyses (called post-processing) for a 3D imaging we did not think about at this stage, but we did not have enough time to do another thing than preparing the near Athens Symposium (September 1988).

Of course it is extremely tedious to hand analyse the numerical results given by a printout of many thousands of numbers and not to use computerized graphical means. We did not have the choice, because the theoretical EDF team was initially fixed at the beginning of the Cheops operation and it was quite impossible to call up other workers which were assigned to other tasks, especially after the previous operation on the unknown King's Chamber which was *wrongly* considered as a media fiasco which stopped the whole Cheops operation (see our important conclusion in p. 38 about the existence of internal walls revealed by the drillings in the corridor to the Queen's Chamber). Perhaps by an ill will, according to Marc Albouy, someone announced in advance the discovery of the unknown King's grave, while an inverse analysis was still in progress. But quite certainly in a hurry, the experimental team decided the drillings at the place suggested by an exploration, thinking about the previous success of the second Solar boat.

The experimental team put too much confidence on the exploration and, backed by French and Egyptian officials, they took the risk of making the drillings in front of the media cameras. Unfortunately, the drilling operation was negative for the unknown tomb. Mathematicians and Computer science specialists knew that there was no simple relation between exploration and inversion especially in 3D. But the harm was done and EDF Directorate stopped the Cheops operation **(12)**.

Fig. 3.14 Three-dimensional meshes: 1 element (the pyramid) for obtaining the mean density, 34 macro-elements, then 2 000 micro-elements for the densitogram (EDF-CPGF)

If the King Chamber operation was negative for everybody, it was not for our research team. We knew the existence of internal walls. We learnt much about the failure of the second operation, which could be simply explained by the word *inadequacy* (between measurements, unknowns and methods of numerical solution).

Results on the Imaging of the Surface Density

The computation of the density can be done by different aforementioned methods and the corresponding results are similar. The numerical solution was quite good, despite the low number of measurements in comparison with the number of unknowns and despite the fact that measurement points were not better distributed on the whole boundary of the pyramid. It was dangerous to work on the faces of the pyramid, because of the presence of sand in the steps, so that we concentrated measurement points on the ridges and nearby. We have previously found that the gravity residual was lower for the heterogeneous pyramid than for the homogeneous one. It was not the ideal solution yet, since there was some small residual at the level of micro-structures. An iterative scheme, making use of the obtained result as a new *a priori* knowledge for a further refined model can be worked out, see [Bui, 1993].

The mean density on the pyramid surface presents some surprising results. First, the mean density in the 34 macro-elements was obtained by solving an over-determined inverse problem which involved much more data than unknowns (rectangular matrix 754 $\times$ 34). The computation showed that the mean density in the macro-element is very low at the top d=1.85 to 1.91 T/m^3, a result confirmed later by a finer solution.

At the South, two blocks present a high density d=2.34 to 2.37 T/m^3, then at the level of the King's Chamber, the density 2.74 T/m^3 was found. The latter number higher than the granite density is clearly a *ghost* (wrong) solution predicted by Mathematicians. It was due to the facts that constant C was taken a too high value and the constraints $\rho \leq \rho_{\text{granite}}$ was not considered in this computation. A finer model solution will satisfy the granite density constraint.

At the Northern side, except the entrance block of mean density 2.50 T/m^3, the density is about 2 T/m^3, as shown in Fig. 3.15. Note that two Southern blocks at the top have an extremely low mean density d=1.85 T/m^3. Now the true density of surface stones is about d=2.30 (limestone). This result means that void represents at least 20% of the volume behind the actual visible stones. We remark that the mean surface density of the Southern blocks at the top is high (red colour, in Fig. 3.17.). Does the low density 20% of the volume behind visible stones correspond to a big cavity behind the surface? This question will be examined in the next Chapter.

Is it the void created by the dissolution of plasters by seeping water, which was more abundant at the top than at the King's Chamber level, or is it something else? One wonders if the Egyptian builders had intentionally introduced spaces between stones for lighten the pyramid. There is nothing which was ignored by the Cheops builders.

We shall discuss this important point on the void. Note that the mean value of density at the 4 top macro-elements is very precise because of too many measurements in the nearby ridges were made around the elements.

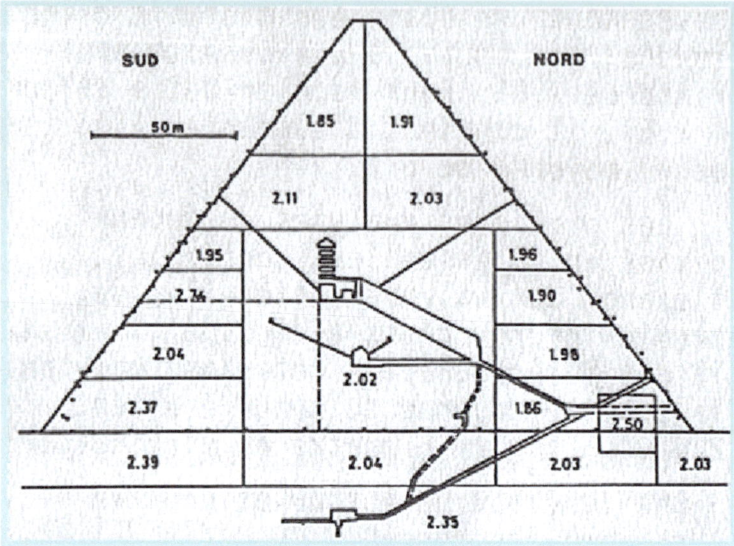

Fig. 3.15 Solution for the finite macro-elements (EDF-CPGF)

After the computation of the one-element model which gave the mean density d=2.05 T/m^3 without any inversion, the updated a priori knowledge $\mathbf{X}_{01}$=2.05 was used for the preliminary 34 macro-elements model, by solving an over-determined problem. Then the density of macro-elements was used as *a priori* knowledge $\mathbf{X}_{02}$ for the finer model of 2 000. We are left with an under-determined inverse problem, with the matrix A (754 $\times$ 2 000) which was solved by the same method. The solution of the first computation of the 2 000 elements model will be used as *a priori* knowledge $\mathbf{X}_{03}$ for a further refined model, *etc*, as mentioned in the 1988 symposium paper. But we had no time to do this because of external reasons (no more budget, change at our Directorate, definitive end of the Cheops pyramid operation, new daily works).

We copy the conclusion of our 1988 Athens Symposium « The Engineering Geology of Ancient Works, Monuments and Historical Sites », also reproduced in Fig. 3.16.

« *Concerning obtained results (the analyses are still in progress) let us mention the most important ones:*
• *The 50 m of the top pyramid are very light (1.85 T/m^3 in mean),*
• *below, heavy zones (2.1 to 2.4 T/m^3) alternate with light ones (1.85 to 2.00 T/m^3), without resulting from a simple symmetry other than a certain spiral shape. Different hypotheses are currently considered in the gravity model, for testing its validity* ».

When we wrote this Athens symposium paper, we made quickly a hand graphics by dispatching on the plane x,y, the points having a surface mean density lower or equal to 1.9 T/m^3. We already quote a « *certain spiral shape* ».

PROCEEDINGS OF AN INTERNATIONAL SYMPOSIUM ORGANIZED BY THE GREEK NATIONAL GROUP OF IAEG / ATHENS / 19-23 SEPTEMBER 1988

Géologie de l'ingénieur appliquée aux travaux anciens, monuments et sites historiques, Marinos & Koukis (réd.)
© 1988 Balkema, Rotterdam. ISBN 90 6191 793 X

Les possibilités offertes par la microgravimétrie dans l'auscultation de monuments antiques

The application of microgravity survey in the endoscopy of ancient monuments

Huy Duong Bui & Jacques Montluçon
Electricité de France, Paris, France

Jacques Lakshmanan & Jean-Claude Erling
Compagnie de Prospec

Chawki Nakhla
Organisation des Antiq

- les influences gravifiques d'un certain nombre d'ensembles subdivisant la pyramide (y compris ...

lors du traitement, qui veulent que les blocs aient tous une taille minimale, ... sur les points

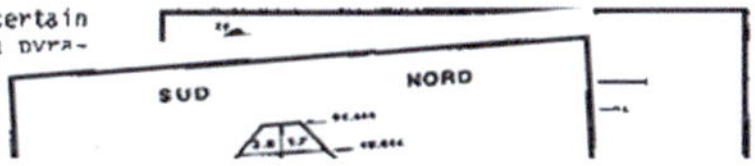

7. CONCLUSIONS

Nous pensons que les prospections microgravimétriques conduites à l'intérieur de la grande Pyramide de KHEOPS ont constitué une "première" dans l'archéologie, non pas tant par la recherche de cavités, que par l'application de la méthode à l'auscultation des structures.

La microgravimétrie s'avère être un outil pour l'auscultation endoscopique d'ouvrages de masse tels que la pyramide, mais aussi de barrages ou de remblais.

On peut évaluer les densités absolues des matériaux, dès lors que les mesures sont faites, sur, autour et à l'intérieur de tels ouvrages ou objets finis. Pour ce faire, les méthodes d'interprétation ont dû évoluer d'une façon très importante, et un logiciel complet et nouveau a été écrit.

Ce retour en arrière, ou "feed back", du monde culturel vers le monde industriel, a déjà eu une première application avec l'endoscopie gravimétrique du tunnel hydro-électrique de la COCHE, dans les Alpes Françaises.

En ce qui concerne les résultats déjà acquis (le traitement est encore en cours), signalons en les résultats les plus importants :

- les 50 m supérieurs de la pyramide sont très légers (1.85 gm/cm^3 en moyenne)
- en dessous, des zones lourdes (2.1 à 2.4 gm/cm^3) alternent avec des zones légères (1.85 à 2.00 gm/cm^3), sans qu'il ressorte de symétrie simple autre qu'une certaine allure en spirale. Diverses hypothèses de construction sont en cours de modélisation gravimétrique, pour en tester la validité.

REFERENCES

Blizkowsky, M., 1979. "Processing and Applications in Microgravity Surveys", Geophysical Prospecting, vol. 27, n° 4 pp 848-861.
Bui, H.D., Lakshmanan, J., Montluçon, J., et Nakhla, C., 1987. "First results of structural analysis of the Cheops Pyramid by Microgravity", Egyptian Antiquities Organization, Int. Symp., Cairo.
Dormion, G., Goidin, J.O., 1986. Kheops : "Nouvelle enquête", Ed. Rech. sur les Civilisations, Paris.
Erling, J.C., Roques, G., 1983. "Reconnaissance et Traitement de Cavités Naturelles ou Artificielles dans le Domaine Ferroviaire", Assoc. Int. Geol. Ing., Symp. Int. sur les Essais en Place, Paris.
LaFehr, T.R., 1983. "Rock density Borehole Gravity Surveys, Geophysics, vol. 48, pp 341-356.
Lakshmanan, J., 1963. "Reconnaissance de cavités dans le sous-sol par procédés électriques et gravimétriques", Sol-Soils, Paris.
Lakshmanan, J., Bichara, M., et Erling, J.C. 1977. "Etudes de Fondation en Terrain Caverneux : place de la gravimétrie", Bull. Liaison du Labo. des P. et Ch., Paris, 92, pp 74-79.
Lakshmanan, J., 1985. Corrections généralisées de Nettleton, Symposium on Potential Fields in Rugged Topography, Lausanne, I.G.L. Bulletin, n° 7, paper 2.5.
Montluçon, J., Lakshmanan, J., "Microgravity probes the Grand Pyramid", Geophysics : the Leading Edge to Exploration, Jan. 1987.
Nettleton, L.L., 1939. "Determination of Density for Reduction of Gravimetric Observations", Geophysics, vol. 5, pp 176-183.
Tarantola, A., 1987. Inverse Problem Theory : Methods for Data Fitting and Model Parameter Estimation, Elsevier, Amsterdam.

Fig. 3.16 Copy of the conclusions of the Athens Symposium (1988) on « The Engineering Geology of Ancient Works, Monuments and Historical Sites », [Bui et al, 1988] *« Below heavy zones (2.1 to 2.4 T/m^3) alternate with light ones (1.85 to 1.95 T/m^3), without resulting from a simple symmetry other than a certain spiral shape. Different hypotheses are currently considered in the gravity inversion model, for testing its validity»*

As mentioned before, we did not think about the rich means of graphical post-analyses of EDF for a automatic display of the density image representing this spiral (in 3D view) and more generally the slice imaging of the pyramid like that of a medical scanner imaging. Even if we had the idea, we could not do that because we had to ask our colleagues of the Computer centre for help in the graphical display matters. We were in summer holidays and the 1988 Athens Symposium was close. The rich computerized graphical means of EDF might be useful for the Cheops pyramid study, as they were for the reconstruction of the puzzle stones of the Akhenaton Talatat [Gondran and Vergnieux, 1997]. So we made an unfinished work when we wrote «*the analyses are still in progress*» in the conclusion of our (1988) paper. And if we had one month more, would we have done? We would give a graphic display on the plane x,y of this square « spiral », obtained just after the Symposium by a careful analysis of the result. The image shown in Fig. 3.17, gives the distribution of the *mean* density of surface micro-elements, of about 10 m thick at the bottom and 3 to 4 m thick at the top.

The Densitogram

The image in Fig. 3.17 of our work has been never published by the author **(13)** in a book or paper. It was published (in black and white) by some authors, who obtained it from our colleagues. The trapezoidal micro-elements have a horizontal size which decreases as we rise in the pyramid. The mean density is defined over the depth of surface element of about 10 m at the bottom and 3 to 4 m at the top.

We recall that it is not the true surface density of stones, which are the visible ones with the density 2.35 T/m^3 of limestone, behind the missing *finishing* stones. It is rather the *homogenized* density over the surface and within a certain *depth*, corresponding to surface micro-elements, which takes into account the void interstices between stones (14). The values in the densitogram are those of trapezoidal shape elements with decreasing thickness as we rise to the top. If one has important interstice voids, the mean density is low (in Green, $d \leq 1.85$), while a compact masonry or simply an arrangement of well squared stones put down without joints, for which the density is clearly higher (in Red, $d > 2.05$). Since the mean density, in a vertical view, is defined for micro-elements within some depth, the interpretation of the densitogram in terms of the pyramid construction is delicate, depending much on the filling of the spaces between the degree walls, the terraces and the pyramid faces. If we assume a perfect homogeneous structure, including the filling of stones as mentioned above, we then have a uniform red colour square. If the pyramid is built like the degrees Djoser pyramid, the filling of the cornices of which is made with low mean density stones, we then have a succession of square green rings on a red background. The discontinuity of square rings can be explained by many ways, but it is difficult to have a more precise idea on them. One may think about the change on the manner to put down stones in the cornices or the presence of a compact

Fig. 3.17 Densitogram giving the density of surface finite elements (Foundation EDF) (Alteration and modification by adds to the image are forbidden)

masonry stair-ramp in the cornice which changes the apparent surface density. A precise analysis of the filling in of the cornices in a degrees pyramid is given the Chapter 5. For example, in the South and the North, from the bottom to 40 m high, high density is observed. We remark a square red zone at the Northern foot which might correspond to massive chevron stones at the entrance. On the other hand, the top of the pyramid is light except its Southern part.

At first sight, especially in the 3D representation of the densitogram given below, it seems that there are more or less horizontal strata or degrees or terraces. The green terraces are connected together by green « bridges » which suggest a « spiral » structure which is explained in the Chapter 5.

At the time we all thought about a construction spiralling up like the Babel tower, not the Babylon tower of the Genesis (11, 1-9) which was a eight degrees monument, with zigzag ramps in opposite directions as shown in the beautiful reconstruction image by Kerisel (1991), but rather the Babel tower in the oil paint by Peter Brüghel at the KHM museum in Vienna (www.khm.at/en/kunsthistorisches-museum/collections/picture-gallery/netherlands-15th-16th-centuries/).

Recent theories are in favour with internal ramps (Houdin's theory) or with inscribed ramps in the cornices analogous to the ramps of the Nabuchodonosor II tower at Babylon. At the time we did not have a precise idea on a coherent interpretation of the results in terms of the pyramid construction. We did not go further in our investigations, to know if the red colour corresponds to the compression of stone ramps or the green colour corresponds to the bad filling in of stones. Utterly perplexed with the problem, we put aside our gravity study and went back to our works on the nuclear power plants.

Fig. 3.18 Babel Tower of P. Brughel (Permission of KH Museum Vienna)

Egyptology was not our line of business and more pressing works were waiting us daily: nuclear vessels, vapour generator tubes, concrete container building, neutron diffusion, vibration of neutron sensors, Fluid-Structure interaction, Computer Aid Design, Artificial Intelligence, Finite Elements Code, especially the *Code_Aster* in Structural Mechanics, *etc.* So a fabulous scientific adventure was ended in 1988.

For lack of discovering a chamber with the King's treasure, we discovered the genius of Egyptian workers who defied the law of gravity to build the first of the seven Wonders of the World, which still keeps and shall keep its mysteries for a long time.

Raising the Density

To obtain a different display of the same results on surface gravity, it is interesting to *raise* the surface density from its vertical projection to the four faces of the pyramid. This operation is the inverse of the vertical projection. The triangular area of the projection of one face with the side a=232.80 m and the height h=a/2, is stretched at it summit so that the new height becomes $H=\varphi h$ (in Mathematics, this operation is called *raising*), where φ is precisely the *golden number* $(1+\sqrt{5})/2=1.618..$, (see Wikipedia) used by Djedi, the magician of Cheops. We then obtain the density on the faces of the pyramid. So far, the golden number is hidden in the Cheops pyramid geometry itself, as previously noted by many authors.

From the new surface gravity, it is easy to give the 3D view of surface gravity density at different angles, Fig. 3.20. The 3D raised image of the South-East, Fig. 3.20, allows us to make a comparison between our results with some theories of the pyramid construction, see also Chapter 4.

Fig. 3.19 Raising the density distribution on the pyramid faces

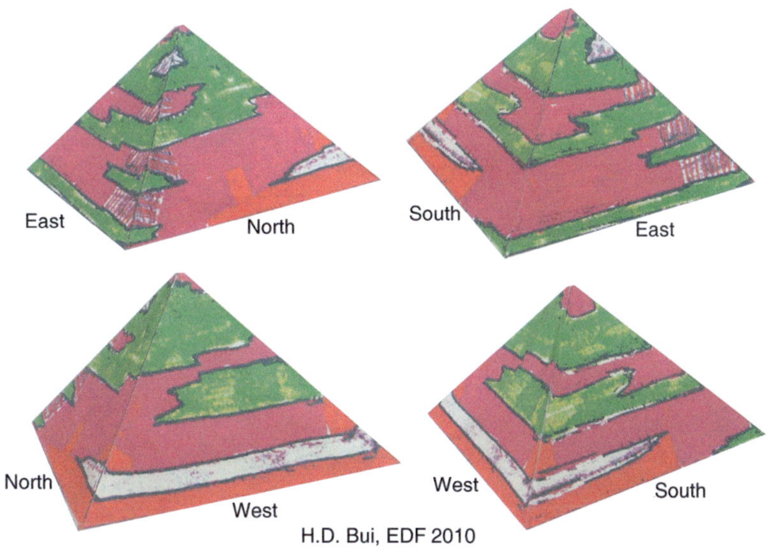

Fig. 3.20 The 3D views of the surface density under different angles

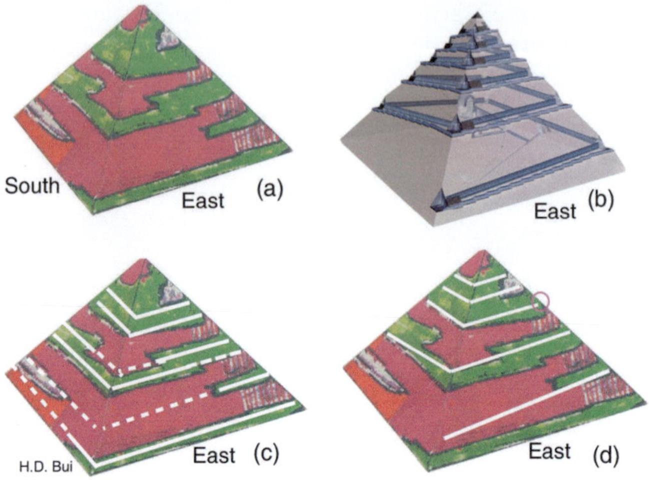

Fig. 3.21 Houdin's internal tunnel or degree theories? (**a**) S-E view, (**b**) Houdin's theory (2000) (Permission of J. P. Houdin), (**c**) Degree theories (Borchardt, Holscher, Guerrier, Dormion) (**d**) The red circle indicates either a Houdin's notch or a platform A or A' of Holscher's zigzag ramps

We see a good agreement between Houdin's theory with our densitogram, Fig. 3.21b and d. Is it the only theory which agrees with the densitogram?

Let us state at once that the *uniqueness* of the inverse problem solution is not ensured, especially because we are concerning mean value of density over a finite micro-element and for that there may be *many* possible arrangements of stones inside the micro-element which have the same mean. Therefore, there is no unique theory of the pyramid construction which agrees with the densitogram.

The question of uniqueness of the theory will be examined in the next chapter where we shall discuss theories with degrees (Borchardt, Holscher, Guerrier, Dormion).

We shall see how theories of degrees also explain the densitogram as well, Fig. 3.21c and Chapter 5.

Chapter 4
Virtual Reconstruction of the Pyramid

It is better to omit some things that may be true, than it is to include a number of dubious theories....
W.M. F. Petrie

There are already sufficient theories on the Cheops pyramid construction by distinguished authors for us to add yet another which was not supported by archaeological, historical or scientific elements. In this chapter, we *review* some theories on the Cheops pyramid construction by restricting ourselves to those that have a connection with our density imaging approach. We shall *not* introduce a new theory of the construction and we shall consider different steps of the construction.

Our aim is to discuss existing theories related to our *real* imaging of the pyramid. After all, with limitations to surface density, we cannot say anything about the inside of the pyramid, apart from the mean density of the pyramid about d=2.05 T/m^3. This mean density was obtained very simply and accurately without any inversion of the gravity equation, except the division between two numbers, from measurements and observable cavities of the pyramid. Theories of internal ramps deeply inside the pyramid given in Kerisel (1991), Rousseau (2001) are not considered for this discussion. We regret that data were deleted from EDF computers 24 years ago and that some documents have been lost since then. With these files or documents, we would be able to get new 3D imaging of the pyramid, for example sliced images similar to medical imaging of our body.

We shall now try to build without justification a *virtual* image of the density according to the Holscher theory of the Cheops pyramid, about which little was said and see how it can be used to interpret our densitogram. Then we make a virtual reconstruction according to Holscher's theory and justify the imaging in Chapter 5. Our limited aim does not allow us discuss unsolved questions about the manner used by the Egyptians to raise stones and the various machines employed in the process. It is up to the historians and the Egyptologists to resolve the mystery, basing their deductions for example on the text of Herodotus who spoke about « *machines made of short pieces of wood*».

On the other hand, we can raise many questions which still make for discussions today and which are related to our results: Were the ramps small or large, internal or external? How were they positioned? What scenario of the construction

H.D. Bui, *Imaging the Cheops Pyramid*, Solid Mechanics and Its Applications 182,
DOI 10.1007/978-94-007-2657-4_4, © Springer Science+Business Media B.V. 2012

outside the well studied central block comprising the King's Chamber, the Queen's Chamber, the Grand Gallery and the corridors? Was the pyramid built « *layer by layer* » « *from the bottom to the top* » as in any classical construction today, or on the contrary, according to a text by Herodotus dated 2000 after Cheops, the pyramid was « *finished off initially at the top, then they moved down to the parts immediately below it and finally added the last touches to the levels closer to the ground and the base of the building*» (Herodotus, *Book II*, 124-124, 127-128, *An account of Egypt*). In short, according to Herodotus, the pyramid was « *finished* » from the top to the bottom. Finally, the question raises about the greatest mystery of the Cheops pyramid, *viz.*, which is "where is the unknown tomb of the King" ?

Our imaging of the density can suggest some direction of research, some models of construction, but we do not propose a new theory of the construction.

Most Egyptologists are generally of opinion that the Cheops pyramid complied with the tradition of the Ancient Empire, even if it innovated in the huge size of this monument. As shown by archaeological remnants, the pyramids were generally built around a core in the form of *degrees*. Some examples are well-known: The six degrees pyramid at Djoser, the unfinished pyramid of Sekhemkhet, the ruins of the Meidoum pyramid, the mastaba of Saqqarah *etc.*

The symbolism of the degrees designed like a giant *stair-case* to allow the spirit of the Pharaoh to rise and reach the other World is confirmed by documents at a later date of the construction, such as the papyrus mentioned in Piankov and Rambowa, Mythological Papyri Texts, (1912), Fig. 4.1 and in Goyon (1977).

Fig. 4.1 Mythologic Papyri (Piankov and Rambowa)

A strange fact that about the Cheops pyramid that there is almost nothing on paper written by Egyptians at the time of the construction, except the name *Khufu*, some signatures on stones by worker teams or quarrymen and a representation of the pyramid with its small top pyramid or "pyramidon". An allusion to the pyramid is found in the Westcar papyrus, exposed at the Berlin Museum, which tells us the story of Djedi, the magician of Cheops, who knew the « *mysteries and secret chambers of the Thot sanctuary* » so that Cheops could make something similar for his horizon (pyramid). We have not yet found any engravings about the Cheops pyramid construction.

The Holscher Ramps and the Steps of the Construction

There are many theories about use of external ramps. G. Goyon (1977) proposed a helicoidal outer ramp as did S. Chapman (2003). Holscher proposed a zigzag ramp on one face, which is an *inscribed* ramp in the pyramid, with possibly a *small* part of its lying outside the pyramid to be removed at the end of the construction. According to Egyptologist G. Goyon (1977), it is the most valuable and rational theory, if the degrees are not too high or the ramp slope is small. Indeed, any ramp *exterior to the future surface* of the Pyramid, requires first the assembly and then the dismantling of stones which are exterior to the pyramid surface, thus involving a huge amount of "double" work. A necessary dismantling of stones must be reduced as much as possible. Why not to imagine that the inscribed ramp was used both to «raise» stones as to «fill in» the cornices, *i.e,*. both works done simultaneously? It was more economic in energy and in time saving.

It is exactly what is proposed by the Holscher model. If it is possible to do that, why not by the Egyptians? A similar plan of construction is found in the Chapman model, with a ramp spiralling up along the degree walls. The Goyon ramp corresponds to the cornices of a helicoidal degree pyramid, like in Babel's tower.

Better yet, let us examine a Holscher model with zigzag ramps in *four* faces, Adam, (1975). It simply inspired by the degrees tower of Babel of Nabuchodonosor II at Babylon (560 BC, thus much later after Cheops), which was made of ramps, but not on all faces, Kerisel, (1991) Fig. 4.2.

Why the number 4? This too has several meanings in Egypt: the Cheops pyramid is directed according to the 4 cardinal points, the number 4 in the rule (3,**4**,5) to get a rectangular triangle by the formula

$$3^2 + 4^2 = 5^2$$

So, the Egyptians likely knew about the rectangular triangle theorem for the construction of the pyramid, 2 000 years before Pythagoras, who was himself the contemporary of the Chinese mathematician Chou Pei, whose graphical proof of the rectangular triangle theorem was very elegant [(15)].

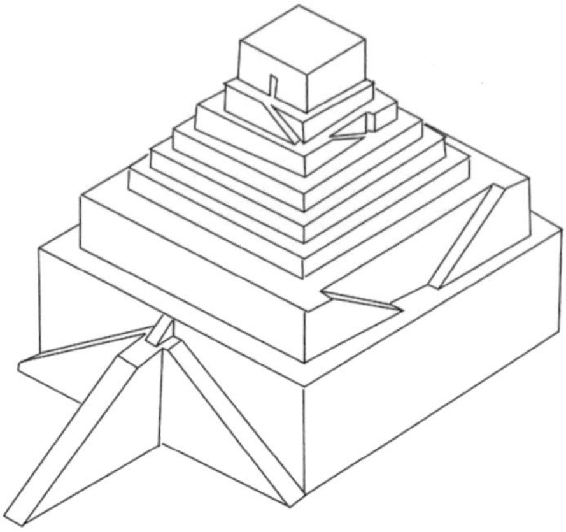

Fig. 4.2 Tower of Babel. Adaptation of the original coloured drawing of Kerisel (1991), Museum Staatliche Berlin. Seven turns are drawn here instead of eight as was mentioned in the text of Herodotus on Babylon

This knowledge which was probably familiar to the magician Djedi allowed the Egyptians to set up the pyramid with perfection. It was an intelligent way to lead a construction on 4 faces, what would allow to divide the duration of construction by 4. Time-saving was important for a construction process which lasted 20 years. Let us imagine a construction once the central core was finished, after the pose of the King's Chamber chevrons. We can imagine several teams working independently, Fig. 4.3:

a) The team who put *very thick puzzle stones* of the Petrie sequence, for the foundation of upper degree masonry,
b) The team for the construction of upper degree masonry and the *inscribed* ramps made by a compact masonry with mortar (red colour),
c) The team of workers who filled in the degree masonry with irregular *quarry stones* of low mean density of the ensemble (green colour) like those visible stones that can be seen through a big hole wall of the Meïdoum pyramid,
d) The team of *land surveyors, layers of corner stones* and *finishing stones* (in Yellow), workers who filled in cornices with *quarry stones*. The precision of the construction depended a lot on the layers of corner stones.
e) workers who *supplied* the building site with *stones & food* and *evacuated rubbles, debris etc.*

So a lot of people and materials called for a huge exterior ramp or several small ramps. A papyrus of the British Museum, called Anastasi I, indicates a frontal ramp of size: 28 m wide (55 cubits, in Egyptian unit), 300 m long (730 cubits), 32 m high (60 cubits). Such a ramp could raise 50% stone volume of the pyramid. We remark that four inscribed zigzag ramps on four faces, of 16 m high in zig and 16 m high in zag, 150 m long each, of 7 m wide, correspond exactly to the capacity of this frontal ramp. It is not excluded that there were other small ramps near the ground for the protection of some finishing stones already put down at the bottom,

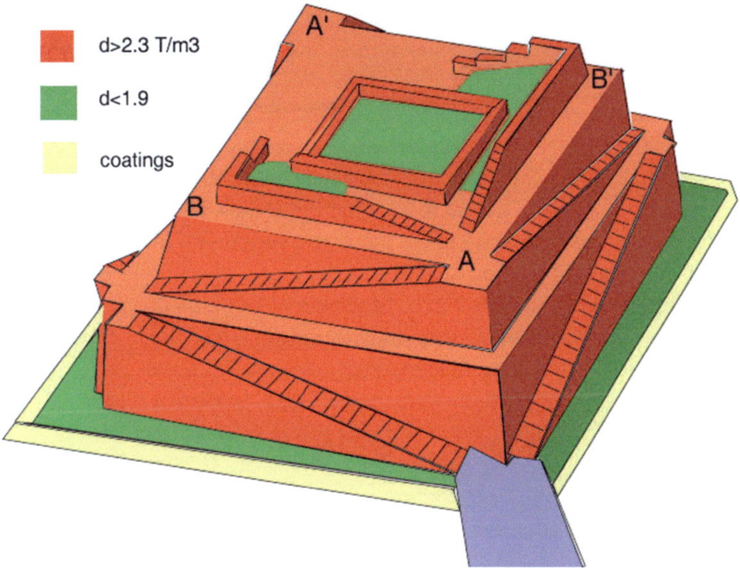

Fig. 4.3 General view of the building after the completion of King's Chamber

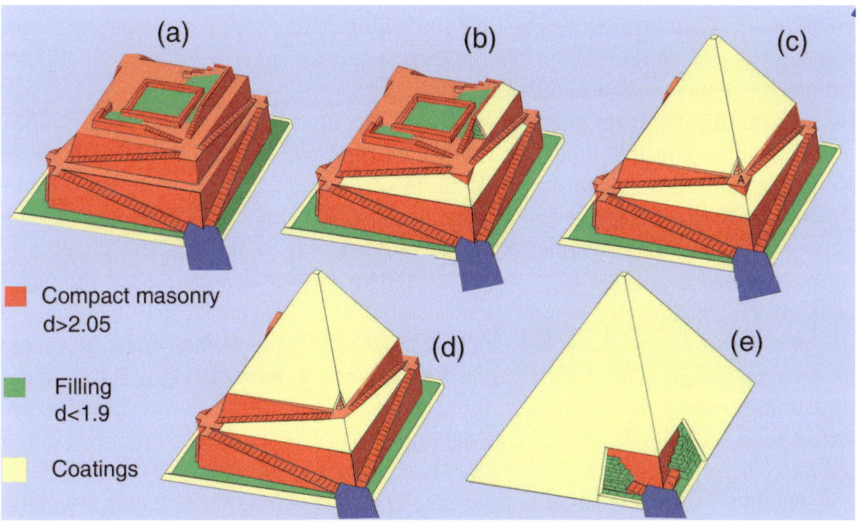

Fig. 4.4 Stages of the construction. (**a**) Construction of degree masonries and inscribed ramps in Red (Borchardt, Holscher); Filling of stones inside the degrees (*Green*). (**b**) Filling of the cornices and/or finishing with coatings (*Yellow*). (**c**) Finishing of the upper part. (**d**) Removal of outgrowth stones of platforms A, A'. (**e**) Backward filling of the cornices (with squared stones from the bottom to the top, in *Green*)

or “internal ramps” and “internal opposite ramps” of Kerisel (1991), or Rousseau‘s ramps (2001). The latter ramps were first in the open air and then covered by a corbelling roof.

A gigantic Borchardt exterior ramp was used by Houdin (2006) to pull granite megaliths weighting 50 to 60 T at the level of the King’s chamber ceiling. But some others did not consider the exterior ramp and pulled all megaliths into the center of the future construction at the beginning of the construction site, Rousseau (2001). They imagined small ramps at the construction site to hoist the megaliths, the “pyramidon”, layer by layer.

The (modified) ramp model of Uvo Holscher (1878–1963) on four faces meets all these objectives. It makes it economically in terms of saving time and energy. It allows the work to be done independently by several teams at the same time. In a single work, two operations were carried out: especially the inscribed ramp construction in cornices being also the filling of the same cornices.

On each degree, Fig. 4.3, we see that the cornices are free of movement of workers and of materials, with the exception of arrival (or departure) platforms A and A’. Figure 4.4b shows the *optional* beginning of the filling in (green colour) and finishing of cornices (yellow colour). However, one would not finish off cornices while they would be used for other purpose. For example, we can envisage that workers camped on free cornices, to avoid going up and going down every day and to be in good condition each morning. One can also imagine that two or three ramps would be of use to raise stones and the remaining ramps would serve on the return of machines and workers. One can envisage one-way street for thousand workers and machines with a minimum police contingency.

Once the top platform was finished - the top small pyramid, the “pyramidon” is not considered in this model. It was called *benbenet* in ancient Egyptian (from Wikipedia), one came down to finish the lower degree Fig. 4.4c and then « backwards » on the ramp as suggested by Egyptologists Goyon (1977), Rousseau (2001), by *removing* some excess stones at platforms A, A’, Fig. 4.4d.

So, to finish a cornice of a degree as shown in Fig. 4.4e one used the usual techniques of construction, namely *stones including rough coating ones were placed one on another, layer by layer, from the bottom to the top of the degree. Then one polished the rough coating stones of each cornice* - Goyon (1977) p. 231 had the same opinion about the finishing of the facade. The most difficult was to realize the perfect joint with finishing stones of the upper degree. But the Egyptians mastered perfectly the laying of bevel-edge stones exactly. It is incomparably less difficult than when they used dovetail granite stones to shut down the sarcophagus in granite of the King and a system of rods falling in holes to block the cover.

In short, it is the scheme used by several eminent Egyptologists, without the small top pyramid. The Holscher model with zigzag ramps in 4 faces allows to have four times more quickly several independent operations going at the same time, without waiting the finishing of stones, with the exception of the ramps which were finished *backwards* at the end, by the « complementary stones » about which Herodotus spoke.

All that seems to be coherent with the text by Herodotus who wrote, 2000 years after Cheops « *The pyramid was built in steps, battlement-wise, as it is called, or, according to others, altar-wise. After laying the stones for the base, they raised the remaining stones to their places by means of machines formed of short wooden planks. The first machine raised them from the ground to the top of the first step. On this there was another machine, which received the stone upon its arrival and conveyed it to the second step, whence a third machine advanced it still higher. Either they had as many machines as there were steps in the pyramid, or possibly they had but a single machine, which, being easily moved, was transferred from tier to tier as the stone rose — both accounts are given and therefore I mention both. The upper portion of the Pyramid was finished first, then the middle and finally the part which was lowest and nearest to the ground.* (16) » *Book II*, 124–127. Translation by George Rawlinson.

[Note: *Tier* means layer of stones (or Petrie sequence); *Step* means degree (or stair-case of the papyrus) made of about 12 tiers or more; according to L. Borchardt, the pyramid contains about 9 steps].

There is no contradiction between real methods of construction and those described by historical texts. On the one hand, stones are put on stones, layer by layer, thus from the bottom to the top of the degree (Herodotus wrote « *tier to tier as the stone rose* ..»). On the other hand, it is nonsense to interpret the Herodotus text as an indication of the construction was pursued from the top to the bottom. He rather said the « finishing » from the top to the bottom, which is not the same thing. Eckart Unterberger expressed the same opinion « *If we assume that the pyramid is a step pyramid, then what Herodotus said would make sense*».

In reality the two points of view are compatible. Inside the pyramid cornices of the degrees, stones were laid layer by layer, stones over stones from the bottom to the top. Workers finished the upper cornices first, according to usual method of laying of stones over stones from the bottom to the top. Then, ones finished lower cornices, from the top cornice, then the middle cornice, to the bottom cornice.

As a matter of fact there were several stages of the construction: the masonries of degrees and inscribed ramps, the filling of cornices and finishing of stones there, except the ramps which were filled and finished backwards from the top to the bottom.

Macroscopic and Microscopic Points of View

Herodotus was right in the operation planning for a degree pyramid and those who contested Herodotus and proposed the laying of stones, including finishing stones, from the bottom to the top, were also right in another context. Heredotus spoke about the pyramid seen by the people from a distance, which was finished from the top to the bottom. His account corresponded to the *macroscopic* view of the construction by far, while others thought about *microscopic* details of the laying of stones seen very near at the cornice level. In reality, the two points of view are complementary (17).

The two complementary aspects have been mentioned in our numerical model when we spoke about macro-elements and micro-elements. In connection with the Egyptian Thot Hermes, Aufrere (2007) also spoke about infinitely great and infinitely small things and about macrocosm and microcosm on religious and mythic levels.

The macroscopic level is coming from the microscopic one through a process called *homogenization*, which is a word used by Mathematicians and Mechanicians. Another word little used is *heterogenization*. When the macroscopic level imposes its unbearable order to the microscopic level, this last reacts violently. It results a heterogenization from it. Witness this beautiful cloud (a macroscopic view) made of water molecules (a microscopic level) becoming a threatening cloud, changing its white colour into a dark one, accumulating the electric charges to the point of launching lightning bolts (heterogenization), which make hear angers of the *God Seth* to be heard by the mortal Egyptians. Another macroscopic order is born with the rain. It is the *phase change* in Physics which characterizes the heterogeneity [(18)].

Here is another example. When the microscopic level does not support any more the mechanical loads imposed by the macroscopic level, it changes the macroscopic level itself, for example earthquake, landslide, instability phenomena (the sliding of added stones to the ancient core of the Meïdoum pyramid), *etc*. They are phenomena that prove very difficult to study and especially to predict.

If the passage of the microscopic level to the macroscopic one is well understood, the inverse process is not. It is easier to solve an inverse problem where the Mathematics are known, than to deal with an inverse process where the Physics are not known.

The Densitogram and the Borchardt Pyramid

The Holscher ramp, which is an open air ramp, is interesting when its slope is small. If there are 9 or 10 degrees in the pyramid, the degree height is about 14 m. The first ramp of 5 m wide has only the slope 7%. One can then use it to raise stones about 2 to 3 tons by sledges. The first three ramps enable to raise half of the volume of stones up to the height 42 m. At middle height, the ramp may be a stair-case and stones are raised by machines "tier by tier". We do not have an exact idea of the machines used to raise stones.

Comparing the degrees pyramid of L. Borchardt (1922) with our densitogram, E. Guerrier (2006) saw the degrees in the Cheops pyramid [(19)]. He would be right. Indeed, the comparison is good for a 9 degrees pyramid, without the foundation. By filling the cornices with more or less joined stones, thus with different colours we obtain an image analogous to the densitogram, Fig. 4.5. It is not a proof yet, but rather a possible explanation.

This *reconstruction* is rather similar to that the judiciary police makes to gather the elements of the facts, not to make evidence of them, but just to allow the judges to forge a conviction. It is not exactly the *reconstruction* in the mathematical sense of inverse problem solutions which implies logical deduction or proof and experimental data fitting. The aim of our discussions is to show that the Borchardt and Holscher ramps are compatible with our densitogram and nothing else.

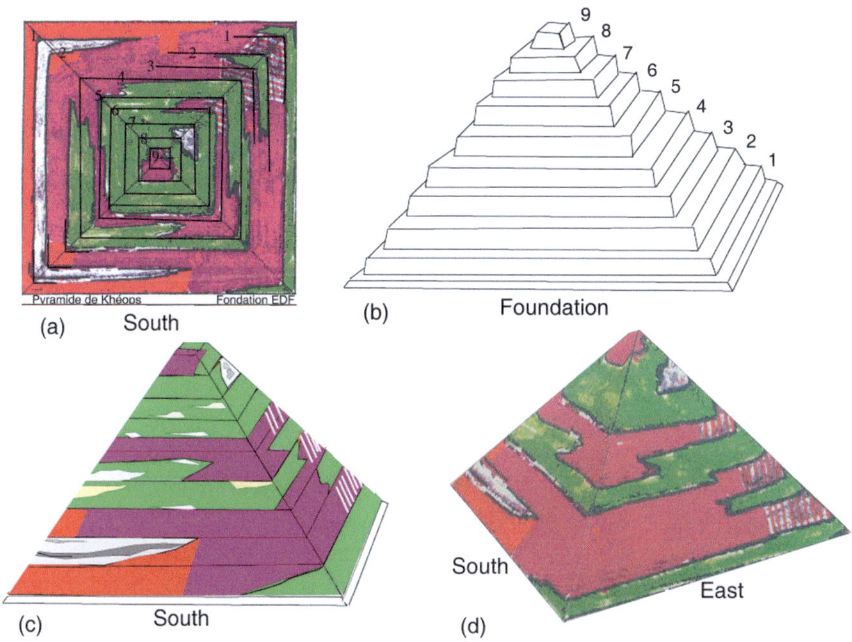

Fig. 4.5 Filling of the degrees (**a**) Vertical projection of the density (**b**) Pyramid with degrees (Borchardt, Holscher, Guerrier etc) (**c**) Reconstruction of the densities by adequate filling of the cornices (**d**) 3D representation density from microgravity on the SE side

Which explanation is there for the arrest of the white band in the middle of the second Southern cornice in the West? Would this be the rising ramp which ended at this place? It is always at bottom of the ramp that there is more stones with light apparent density because of gaps, while the top of the ramp belongs to the masonry of the upper cornice, consequently there is a transition from the green colour to the red tone (see Chapter 5).

Why is there a green band at the Eastern base? Could this be the trace of a ramp of small slope parallel to the Eastern base?

Why is there a green colour at the South-Eastern corner? Is this related to the fact that many stones in this place are missing? Without any doubt, because we have many stones which are missing on this South-Eastern edge on the level of the ground, see Fig. 3.8 (Tatiana's photo). Our numerical computation made use of a perfect geometry so that there are so many voids in the finite element at this place.

Another green trace at the Southern foot which ends shortly indicated a steeper ramp going up towards the West about 50 m.

On Fig. 4.4c, at the fifth cornice, one sees two green bands which would be the trace of two rising ramps arriving at the same platform A.. The stones of the platform A of the cornice would belong to masonry and would be of high density. On vertical projection, Fig. 4.4a at the North-Western corner, one finds the same figure corresponding to the symmetrical arrival platform A'.

The Houdin Internal Ramp Tunnel

« *The Pyramid is built not from the outside, but by the inside*», Henri Houdin (1999). It is by this intuition that J.P. Houdin started his theory. In 2000 in Paris, the Houdin (father, Henri and son, Jean-Pierre), presented their new theory of the Cheops pyramid construction which included some novel ideas [(20)].

Two, among the ideas of the Houdin, seem very important to us.

The first idea of J.P. Houdin relates to the Grand gallery seen like a guide of counterweight to hoist the megaliths of more than 60 tons, placed on the gigantic ramp at the South face and at the same time like a store of the large blocks of granite intended to block the ascending corridor and incidentally like a store room to discharge the small stones from the carriage from counterweight to start again a new cycle. A similar idea of guide of an oblique elevator was proposed by P. Crozat (2002), who rather saw the megaliths placed on the Northern side, what implies cables and supports for horizontal traction forces of more than 100 tons. The proposal of J.P. Houdin is technically more satisfactory because the weights and counterweights balance mutually and the resultant of the forces is vertical while in the case of the oblique elevator of P. Crozat, the resultant of the forces is horizontal. In both theories, it would have to be specified how the stones of counterweight were hoisted.

The second idea is the internal ramps parallel to the facades, arriving at « right angles » at « notches » on the edges, some still visible. The internal ramp, like a spiral square staircase, was used to hoist stones above 40 m. This idea is interesting because all the edges of the current level are then free of circulation and the facades can be finished there, by laying stones on stones, layers by layers. If one neglects the finishing of the notches, the pyramid according to J.P. Houdin is seen *from a distance* as if it had been finished *from the bottom to the top*, maybe in disagreement with the Herodotus text. Only if the notches are completed from the top to the bottom, that it makes Houdin say that his theory respects the Herodotus text as well. We have seen that the cornices of the Holscher model are also free of circulation and can be finished, except the ramps which are finished backwards.

In Houdin's theory, it does not seem that there are degrees, in any case J.P. Houdin did not say it and did not show it in his film, www.3ds.com/khufu. But nothing prevents us from imaging that his virtual construction can include degrees too, which are deeply inside the pyramid to appear in our densitogram.

While being rolled up around the pyramid, in the anticlockwise sense (laevogyrous spiral), the Houdin spiral, Fig. 4.6b et c, occupies about the place of the zones coloured in green of the densitogram. By way of comparison, the theories of degrees of Holscher, Guerrier, *etc.* are represented by white horizontal lines, Fig. 4.6d. We indicated on Fig. 4.6c by a red circle the notch visible currently, at 87 m in height. Recently, Bob Brier, eminent Egyptologist and specialist of mummies, got on this platform and discovered a cavity, visited already at the 19th century by G.A.F. Fitzclarence, a cavity high enough that he stand upright (Houdin, 2009).

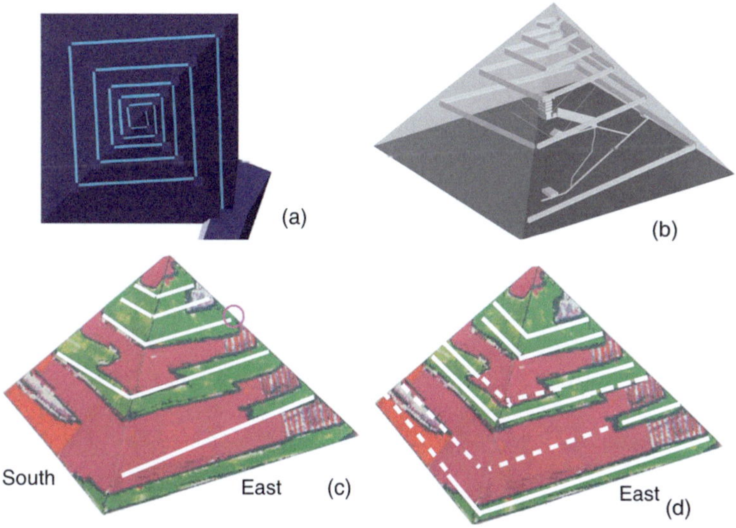

Fig. 4.6 Spirals or layers. The densitogram is reproduced by the SE view. (**a**) and (**b**) the internal tunnel of J.P. Houdin, http://construire-la-grande-pyramide.fr; (**c**) the layout of the internal tunnel of Houdin. The *red circle* indicates the position of the notch visible at 87 m; (**d**) Comparison with theories of degrees

Is it the notch of the Houdin internal ramp ? Or are they the platforms of types A or B of the Holscher with four faces ramps? The two assumptions seem to hold. We do not privilege any theory at the expense of the other. We do not come to a conclusion about the point of knowing which of the two theories is more probable and nearer to our results of measurements and numerical computations, leaving the scientific research of the ones and others to be confirmed (or invalidated) by new discoveries on the site or by papyri and historical texts. Some experts believe more and more in the degrees theory seeing our densitogram as their confirmation (21).

We repeat that the inverse problem that we studied does not have a unique solution. Moreover, as it was a single series of measurements and one calculation, we had not studied the influence of uncertainties of measurement and computation (20). In addition, the densitogram gives the average surface density on a certain depth, two arrangements of stones in the cornice having the same average density in the surface finite elements can exist, what leads to two different theories. There is no either uniqueness of theories of the pyramid construction.

The Mystery of the King Tomb

Does the Pharaoh still sleep in an island surrounded by water under his Pyramid, or beside his pyramid, about which Herodotus spoke? Or does he always sleep in an unknown tomb with all his inviolate treasures in the pyramid itself? It is the greatest mystery of the Cheops pyramid.

It is difficult to speak about the Cheops pyramid without speaking about this mystery, insofar as all our three studies on the pyramid, made in 1986-1987, precisely started starting from this question raised per G. Dormion and J-P. Goidin with us: can one find the unknown chamber? We had explored the track of the cracks of the granite beams of the current King's Chamber, but without success. There are too many assumptions in the geo-mechanical study so that the conclusions are not clear enough. Then we had implemented the Imaging of the density by microgravity. Thereafter, G. Dormion (2004) carefully studied the pavement of the Queen's Chamber and concluded with the possibility of unknown harrows next to an unknown room located a little to the West under the Queen's Chamber. Two other researchers, J. Bardot and F. Darmon (2006), thought of locating the unknown room between the King's Chamber and the Queen's Chamber, while being based on our geo-mechanical results [Montlucon *et al*, 1987] and their discovery of true false-joints at the horizontal corridor.

What does our densitogram reveal? One notices a red spot at the summit platform, corresponding to heavy stones, shifted a little towards the South, Fig. 4.6c, d. The green traces beside it would be logically fillings of one or two staircases for the transport of the stones until the top by low mean density stones with many voids. The dominant colour at the top is green, except the red spot.

What can such heavy stones hide then? It is thought that the King was aware of the cracking of the ceiling of his Chamber, just when the beams had just been posed. It was the opinion of Egyptologist J.P. Lauer. Some other Egyptologists thought that this moment came much later when the pyramid reached level 100 m. The architects would then decide to build a new tomb almost at the top of the pyramid, shifted a little towards the South in order to have room for building ramps and hoisting stones. Since the megaliths solution would be impossible, smaller granite stones would be posed in corbelling, as it was carried out in the Grand Gallery. Access passages to the stores as well as the corridor would open (or not) on slopes or cornices still with open sky. Indices of strong surface density in the South of the summit block seem to suggest the presence of heavy mass near the surface which compensated the void of a chamber. The red colour would be the heavy Southern wall of the unknown tomb. Now according to Fig. 3.15, the average density in the whole summit Southern block is 1.85 T/m^3 (in Green). But on the Southern surface at the summit, the density is at least equal to 2 T/m^3 (in Red). It results from it that inside the Southern summit block, there is a very important vacuum. Is this vacuum a cavity? Could this be the tomb of the King? Maybe one of the last open air tomb evoked in Chapter 1? Not only one summit tomb would be compatible with our microgravity results, but it would be completely worthy of the King.

Let we *dream* with this possibility there. What could be greater than to thus reserve for the King his tomb for Eternity at the top of his Pyramid!

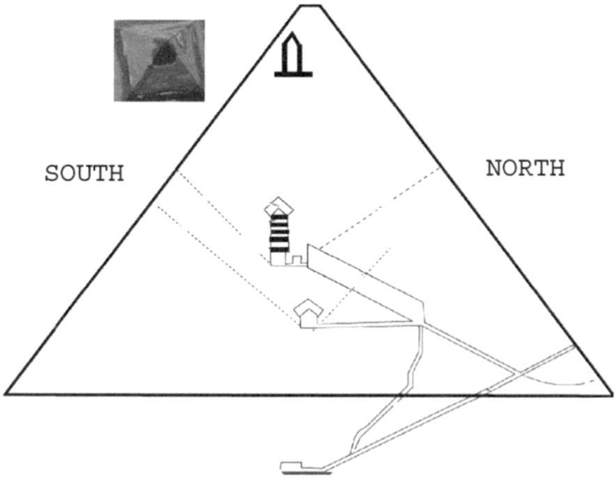

Fig. 4.7 In medallion: 3D view of the densities of the summit area. The heavy zone is shifted towards the South. *Green colour* could be the filling of the staircases towards the summit platform. Below the *red spot*, there is many vacuums. The average density at the top is only of approximately 1.9

Golden Number and Intertwined Spirals

The trace of Houdin's spirals on a vertical view shows a light rotation of a few degrees, in the anticlockwise direction, Fig. 4.5a. The reason is that this trace follows a straight line of the internal ramp between two levels of the Pyramid, which is inclined with respect to the foot edge. The Holscher ramps in themselves are also inclined but alternatively in two directions. As the ramp is inscribed in the cornice volume and that the latter is then filled with stones, with more or less voids and thus with different lower density, it is this cornice that one sees in the measurement of microgravity. For that, the images of the densitogram are parallel to the axes of the coordinates. These remarks show that the densitogram can be explained by the two theories. None is privileged by our result. When displayed equally, spirals appear in a squared shape. When a part of one spiral is missing, the image resembles the another spiral. Let us consider a ramp occupying a cornice. The passage at the higher step is close to an edge of the pyramid. Continuing to go up in the same direction, on another face, one obtains a spiral with line segments whirling around the pyramid. One realizes that there are four distinct spirals, two clockwise spiral (dextro), two anticlockwise (levorotatory). These four spirals intertwining upwards could be seen as an *arabesque*, which goes up to the sky. On Fig. 4.7, we try to put spirals going two opposite directions on our densitogram and we also obtain an arabesque.

The arabesque is incontestably a sign of spirituality which fits in well with the site of the Cheops pyramid. It recalls the spirals in the vegetable world such as for example in the *sunflower*. In the flower, there are however much more spirals, 21 in the clockwise direction and 34 in the anticlockwise one, which are two figures of the Fibonacci sequence 0,1,1,2,3,5,8,13,21,34.., $x_n, x_{n+1}, x_{n+2}(=x_{n+1}+x_n)$. In cone, there are 8 spirals in the clockwise direction and 13 in the anticlockwise one, with

Fig. 4.8 Arabesque of clockwise and anticlockwise spirals

Fig. 4.9 Spirals of the sunflower

figures (8,13) being in the same sequence. For n tending to infinity the ratio x_{n+1}/x_n tends towards a limit which is the *golden number* $\varphi=(1+\sqrt{5})/2=1{,}6180\ldots$The golden number is hidden in the geometry of the Cheops pyramid, whose faces form the angle 51°51' with the horizontal plane, since cos(51°51')=1. Thus the magician Djedi has left nothing to chance. All was ordered perfectly in the magic proportions of the golden section. Here we have only four spirals in the two directions, two spirals in each direction. Arabesques which goes up to the sky and sunflowers which "turns towards the Sun "are both the perfectly adequate symbols to be associated with the tomb of the King. A painting offered to the author by JOT represents an arabesque of intertwined spirals, Fig. 4.10. If we gather now arabesque branches of each face together, we obtain the zigzag ramp of Holscher in the four faces, Fig. 4.11. Chapter 5 describes the way in which are obtained theoretical zigzags, by considering stone fillings of the cornices.

Fig. 4.10 A painting of an arabesque (Permission of JOT)

Fig. 4.11 Regrouping in zigzag ramps of Holscher

Chapter 5
Filling the Cornices

We study a model comprising 9 degrees, for a height of the summit platform with 135 m, by removing 1 m for the foundation raft. The height of the wall of each degree is about 135/9=15 m. It is known that the walls of the degrees of the pyramids have a strong slope of 75° and the face has a slope of 52° approximately. These figures give an idea of the space of the cornice where one will install a ramp in solid masonry (d=2.05 to 2.3, Red colour) and will fill the remainder of the cornice of stones with many voids or low mean density (d<1.9, green colour).

The vertical projection of the inclined wall has a width of L=15/tan(75°)=4 m, while the cornice width is 15/tan(52°) −4=7.72 m or roughly 8 m. It is largely sufficient, because it is question in chapter IV of ramps width of 7 m. But we will not use all the cornice. Indeed, a foot of ramp occupying the totality of the cornice width obliges to have many stones outgrowths at the top of the ramp, stones which should be removed later. A not very broad ramp is insufficient for transport. The Egyptians thus knew the optimal width of the ramp width, but we do not know it.

It is supposed that all the cornice ramps are filled with dense stones, *i.e.* solid masonry stones with mortar, for the stability reason of transport. It will appear in Red in the restored image.

On the other hand if the width of the masonry ramp is of 3 m or 4 m, the filling of cornices with stones to complete the pyramid is not a very dense material, one thus has a portion of the cornice in Green.

True Density and Mean Density

The average density observed on the surface is the average on a certain depth, between current external surface and a surface about parallel with it, distant of D=10 m in bottom of the pyramid and of D=3 m at the top. In what follows we take a means value D about 5 m. To study the average density, let us consider the true density in a volume of reference bounded by the current outer surface, a surface with the depth D parallel with it and two surfaces orthogonal to the outside. The section of this volume in a point of the cornice is studied for the filling.

H.D. Bui, *Imaging the Cheops Pyramid*, Solid Mechanics and Its Applications 182, DOI 10.1007/978-94-007-2657-4_5, © Springer Science+Business Media B.V. 2012

Filling the cornice, according to the case leads to very different local densities. To simplify the presentation, we use the green colour to indicate the filling with many interstices, with density approximately d = 1.9 and the red colour for compact masonry.

One must take account also solid masses of the degree considered and degree in the lower part located in the volume of reference which corresponds to 3 finite macro-elements approximately. The position of the macro-elements in the volume of reference is only indicative, because we do not have any more their exact position in our imaging data. One obtains the result in which the restored image varies from the Green in bottom of the cornice to the Red in top of the cornice. Figure 5.1 indicates the true density to the foot of the ramp in the volume of reference considered and the densities of the restored image indicated by coloured small square medallions. Figure 5.2 indicates the same densities in top of the ramp.

In addition the restored density given in the image is made in a discrete way in some ranges of values, which does not enable us to restore all the small nuances of colour of the average on the depth, which must vary continuously. In Fig. 5.1, it is represented in medallions by two green squares and one red square. In Fig. 5.2 in top of the slope, the proportion is reversed. Thus along the ramp, *a change of colour from the Green to the Red indicates a rise in the ramp*. We then retain this simple rule of the Green-Red transition to interpret our image. But this rule does not apply systematically to the pyramid according to whether the filling is carried out with more or less vacuums.

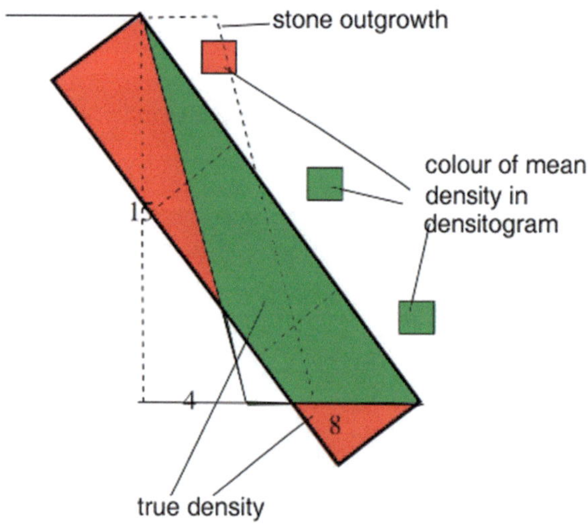

Fig. 5.1 True density and average density at the foot of ramp of 3 m width. Solid masonry in *Red* and filling stones in *Green*. Restored density on the real image is discrete and appears in medallions: *green squares* (density < 1.85) for two macro-elements, red square at the top (density > 2.05). The stone outgrowth in *dotted line* is ignored

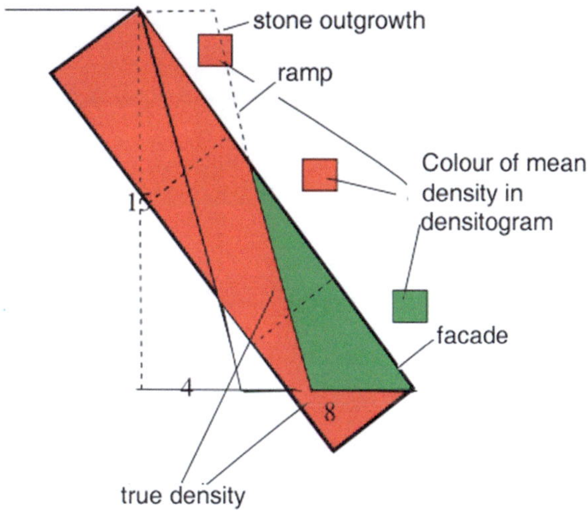

Fig. 5.2 True density and average density at the top of the ramp of 3 m width. There is more red colour in the top than in the bottom of the ramp

Comparisons with Observations

Sometimes, the transition Green-Red is missing. That can explain the absence of transition Green-Red in low parts of cornices. When two images of transition Green-Red are with a right angle one has the *signature* of a platform of arrival of the type A, A' having two rising ramps, Fig. 5.3. It is what one sees on the South-Eastern edge, at the middle height of the pyramid. It is what one also sees, but with a little more uncertainty, on the symmetrical platform A'

When two ramps are side by side and rising in the opposite directions, we have the image in Fig. 5.4. On the left, two theoretical traces of Green-Red transition and on the right, the real image observed on the Eastern face at mid height.

From these approximate rules of the transition from Green to Red, which do not rest rigorously on the calculation of average density in the cornice, we can identify some *black* arrows of the rising ramps in Fig. 5.5. One clearly sees there the outlines of the zigzag ramps of Holscher on the 4 faces. However, by considering symmetries existing probably between the axes North West-South East and North East-South West, we can complete black arrows by white ones for rising ramps in the symmetric part of the pyramid. We obtain thus practically the arrows of the 4 rising ramps in zigzag of Holscher on the 4 faces. That shows well the coherence of the assumption of the 4 ramps in zigzag in agreement with our imaging of density.

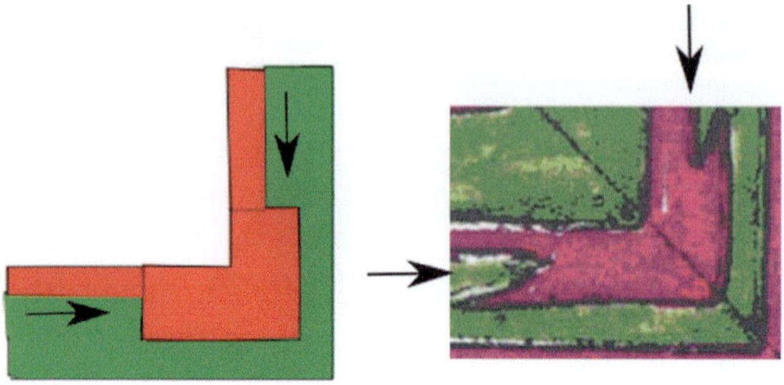

Fig. 5.3 Arrival platform A of two rising ramps. On the left, the theoretical image and on the right, the image at the South-Eastern edge at mid height

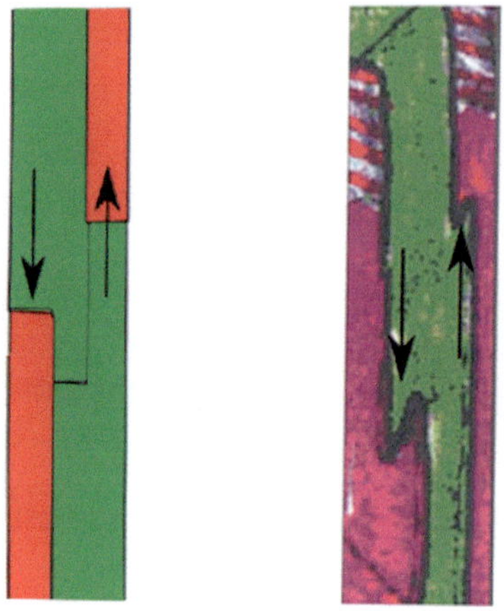

Fig. 5.4 Transition from *Green* to *Red* in two adjacent lines, at 50 m high

Fig. 5.5 Ramps in zigzag. *Black arrows* suggested by the densitogram. *White arrows* completed by symmetry

Notes

(1). Solid mechanics

In Mechanics, one calls beam any solid structure definitely longer than broad. It generally rests on fixed supports at its two ends, supports free or embedded, or embedded only on one support in the case of the brackets. It is laterally charged by pressures which can be null, concentrated or with continuous loadings or couples at the ends.

(2). The graviton

The *photon* carries electromagnetic forces, the *gluon* carries nuclear forces while the *bosons* (in particular the undiscovered *Higg's boson*) and *fermions* carry weak forces in the Standard model. There are other particles in the SUper SYmmetry model SUSY introduced by Pierre Fayet, like *photinos, gluinos, selectrons* which are the partner particles of photons, gluons, electrons respectively. One envisages the existence of the *graviton* which carries the force of gravity. The theories of Newton and Einstein and their generalizations are still far from a unified theory of Physics.

(3). The gravimeter with cold atoms

The gravimeter with cold atoms of the LNE-SYRTE, resulting from works of modern Physics of the Nobel Prize C. Cohen-Tannoudji, reaches a much higher degree of accuracy.

(4). Stochastic inversion method

Marc Bonnet and Xavier Chateau of Ecole Polytechnique wrote a program according to the stochastic method of Tarantola, used by J. Lakshmanan in his thesis. Thereafter, in our study, the program was completely modified to include the constraints of inequalities on densities and especially to introduce the adaptive aspect of calculations. This initially comprised only one element to obtain the average density, then elements of increasingly large number to update the *a priori* knowledge X_0.

H.D. Bui, *Imaging the Cheops Pyramid*, Solid Mechanics and Its Applications 182,
DOI 10.1007/978-94-007-2657-4, © Springer Science+Business Media B.V. 2012

(5). Stability of constructions

If one provides Prof. J. Salençon of Ecole Polytechnique with the geometry and mechanical characteristics of the stones added to the initial core degree of the pyramid of Meïdoum, he could specify the stability conditions of the pyramid and the slip surface which led the pyramid to its partial state of ruin as we know it today.

(6). Radars and Maxwell equations

Maxwell equations for studying radars are vectorial and are much more difficult to solve than the simpler scalar Newton gravity equation.

(7). Structural and functional tomographies

X ray and Gamma ray tomographies constitute what is called *structural tomography*, to study the *structure* of the object which is its density distribution. Other recent methods using the emissions of electrons by biological processes are called *functional tomography*, for example PET (Positon Emission Tomography), SPECT (Single Photon Emission Computerized Tomography). The simultaneous use of the structural tomography and the functional tomography in medical imaging makes it possible to know the positions and *functions* of the bodies. The two teams of the author at EDF (Stephane Andrieux, Amel Ben Abda today at the University of Tunis) and at Ecole Polytechnique (Andrei Constantinescu, Hubert Maigre, Stephanie Chaillat, Eva Grasso), obtained mathematical results for the detection of defects and cracks in acoustic, elastic, elastodynamic, thermal and viscoelastic tomographies (human tissue is viscoelastic). One understands the importance of the studies of inverse problems for the problems of maintenance of EDF power plants. The author also collaborated with K. Mai Nguyen and T.T. Truong of the University of Cergy-Pontoise in medical imaging, involving the conical Radon transform, see [Bui, 2006].

(8). Stacking of spheres

It is the conjecture of Kepler (1611) saying that the optimal density of a stacking of spheres is lower than $\pi/\sqrt{18}$ or 0.74 (26% of voids). This conjecture was proved by Tom Hales of Pittsburgh University.

(9). Instability and reinforcement

According to Kerisel (1991), the slip could be provoked by the earthquake. In fact, there are elements of weakness to explain the lack of stability of the work: great height of the structures added to the initial degrees, smooth faces of the degrees inclined almost with the vertical and thus the induced problems of buckling, mechanical absence of connection between the degrees. It is not the case of the Cheops pyramid, because the thick stones of the Petrie sequence would be posed the ones in complementary with the others under the degree walls like a puzzle and thereby establishing a solid three-dimensional reinforcement between the degrees.

Reinforcements of the core of the Senostris II (1897-1878 BC) pyramid at Fayum by visible vertical masonries walls along diagonals prevented its instability. However added parts were likely collapsed by the same mechanism as that of the Meidoum pyramid.

(10). Exploration and inversion

The link between Exploration and Inversion is very simple to establish when the influence matrix A is a square and diagonal one. It is not the case for gravity equation since matrix A comes from the Volterra integral equation of the first kind, which is non diagonal and very badly conditioned as it is known in Applied Mathematics.

(11). Dormion and Goidin (1986)

Their hypotheses were the stating point of our works on the Cheops pyramid when they asked EDF in 1986 for its validation by gravity measurements. One hypothesis was a new corridor from the Northern Entrance to hypothetical Anti-chambers and a tomb right next the King's Chamber and the Great Gallery We did not confirm the precise location of hypothetical rooms and corridors, because our 2D inversion of the gravity equation, using measurements roughly in a plane, was unable to detect a 3D structure. Dormion (2004) envisaged another funeral complex below the Queen's Chamber with access through a corridor to the East. Our successive "open air tombs" are variants of these funeral complexes.

(12). The Paris meeting in October 1986

In October 23, 1986, a meeting in Paris was organized by EDF after the media fiasco of the second Cheops operation (King's Chamber), under the presidence of Prof. J. Kerisel. The EDF experimental team and CPGF presented positive results on the Exploration of the second solar boat, the negative results related to the unknown King's Chamber, the drillings in the corridors to the Queen's Chamber which found nothing but yellow sands. The media noticed the absence of the unknown tomb. But the discoveries of internal walls by the drillings were unnoticed. The author was not invited to this meeting, since his works really started in 1987, with the *positive* blind test obtained in March 30, 1987. The author thinks perhaps that this positive test saved for a time the third Cheops project continued by our group, strongly motivated by the request of Dr Kadry, whereas the project was not supported any more by EDF.

(13). Henri Houdin, Jean-Pierre Houdin, Eric Guerrier and Aline Kiner

In July 2000, the densitogram was shown to Henri Houdin for illustrating J.P. Houdin's theory of the internal ramp tunnel. Journalist Aline Kiner wrote in her paper in *Sciences et Avenirs,* April 2007 : « *H.D. Bui is said very interested by the theory of J.P. Houdin*», but « *we do not have enough measurements to draw the conclusions, he said and it would be necessary to remake measurements on all the faces* ». Recently, in our correspondences with Eric Guerrier, the defender of the theory of degrees, we did rule out the theory of degrees which could lead to the same images.

(14). Conclusions by J. Kerisel

In the Paris meeting of October 1986, Prof. J. Kerisel concluded: «*The quality of the external stone coatings does not correspond at all to that of the internal volume of the pyramid, as the average density much lower d=2.05 shows it and that one could not imagine it. Behind a thick and dense skin, internal volume would comprise tender stones, below of the scale of the density of the stones, with perhaps a multiplicity of small vacuums in the assembly of the stones which were roughly squared to save labour, which is rather reasonable in comparison with the brevity of the life of the Pharaoh and the wear of the copper saws used for the cutting of the stones. Second assumption, it would comprise cavities embanked by sand. Third assumption, it would comprise a heterogeneous structure, laminated probably, without these assumptions being exclusive one of the other* ».

(15). The Pythagoras theorem, Pei theorem or Djedi?

The Pythagoras theorem was known a long time ago before Pythagoras. A babylonian cuneiform tablet of the Columbia University (Plimpton collection N°322) indicated some Pythagoras triplets (3,4,5), (5,12,13) *etc.*, 500 years after Cheops. The magician Djedi of Cheops would have some knowledge on the theorem used for his perfect construction of the pyramid. The Pythagoras theorem was also discovered by Chou Pei, 2000 years after the Cheops pyramid. The graphical proof of the rectangular triangle property by Pei was used as the logo of the World Mathematical Congress in Beijing 2002.

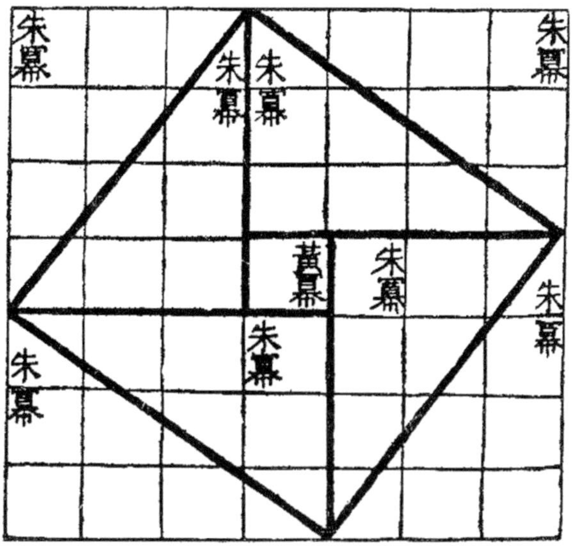

Chou Pei's proof of the rectangular triangle theorem.

Another I.N. Pei is the Architect who built the glass pyramid of the Louvre Paris. The name *Pei* is written in Chinese character as 裴. It was the name of a town in Central China. In 580 BC, Emperor Yu Shun, a contemporary of Nabuchodonosor II, who built the Babel tower at Babylon, rewarded an inhabitant of Pei – we are unaware of his true name, perhaps an important person or a minister - for his service by offering the city to him. Later, part of its descendants emigrated in Canton and changed their name into Pei. However descendants living this area of Canton deformed their names as "Pui" or "Bei" and then by another deformation in "Bui". All these names had the same origin since they used the same Chinese character.

The Pei's glass pyramid of the Louvre Museum (Paris).

(16). Herodotus

The English translation of the phrase is: *The upper portion of the pyramid was finished first, then the middle and finally the part which was lowest and nearest to the ground.*

(17). Complementarity

This aspect is discussed in the paper [Bui and Gondran, 2010].

(18). Frustration in physics and homogenization

There exists a phenomenon in physics called *frustration*. One thing and its opposite or rather its complement can cohabit in the same place. The best known example is ice cream (sorbet) which refreshes us in summer. It is a mixture of sweet water and ice which makes it possible to observe the presence of the two incompatible phenomena of freezing (or *finishing from the top to the bottom*, according to Herodotus) and fusion (or *construction layers by layers, from the bottom to the top*). Those who see only water in a supercooled state, think that the others, who see rather solid crystals, are mistaken. Actually, there are both solid crystals and liquid phases. The two incompatible phenomena lead to frustration, which result in homogenization where both phases are simultaneously present in the macroscopic scale and separated only in the microscopic scale. At the macroscopic scale one has the sorbet, while at the microscopic scale one has two separate phases, crystals and liquid phases.

(19). Possibility of degrees

In a correspondence with E. Guerrier in 2009, the author told him that the degrees theory can lead to the same densitogram.

(20). Sciences & Avenir, April 2007

Journalist Aline Kiner wrote in her paper in *Sciences et Avenirs* April 2007 : « *H.D. Bui is said to be very interested by the theory of J.P. Houdin*», but « *we do not have enough measurements to draw the conclusions, he said and it would be necessary to remake measurements on all the faces* ».

(21). Sciences & Avenir, January-February 2011

About the theory of degrees, journalist Herve Ponchelet wrote in page 50 « *The measurement data on microgravity realized by EDF in 1986 and the numerical interpretations by H.D. Bui of CNRS, can be understood in this sense* ».

In page 14 of the February Issue, Journalist Aline Kiner wrote "*The "spiral" fits well with the internal tunnel ramp of Houdin, even if, today, H.D. Bui proposes another interpretation: according to him, the spiral would suggest a pyramid first built by degrees, then by adding masonry inclined ramp, with the complementary volume filled of stones with (interstices) voids*".

Bibliography

Adam JP (1975) L'archeologie devant l'imposture, Laffont, Paris.

Albouy M (1994) Du Titanic à Karnac. L'aventure du Mecenat Technologique. Dunod, Paris

Aufrere S (2007) Thot Hermes, L'Egyptien, de l'infiniment grand à l'infiniment petit. Collection Kubaba, Editions L'Harmattan

Bardot J, Darmon F (2006) La Grande Pyramide de Kheops. Nouvelles decouvertes. Preface de Z. Hawass. Ed. du Rocher

Borchardt L (1922) Gegen die Zahlenmystik an der grossen Pyramide bei Gise. Berlin, Behrend

Bui HD (1993a) Introduction aux problèmes inverses en mecanique des materiaux. Eyrolles, Paris

Bui HD (1993b) Inverse problems in the mechanics of materials: an introduction. In French (Eyrolles, Paris), English (CRC Press, Boca Raton), Japanese (Shokabo, Tokyo), Russian (Novosibirsk Univ. Press), Chinese (Harbin)

Bui HD (1996) La mariage de la mecanique et des materiaux. La Vie des Sci t 13(5):403–407

Bui HD (2006) Fracture mechanics. Inverse problems and solutions. Springer. Russian translation by Fizmalit (2010)

Bui HD, Gondran M (2011) On complementary theories of the Cheops pyramid construction. Duality, symmetry and complementary in solids mechanics, Ehrlacher A, Markenscoff X (eds). Presses des Ponts & Chaussees, Paris

Bui HD, Lakshmanan J, Montluçon J, Erling J, Nakhla C (1988) The application of microgravity survey in the endoscopy of ancient monuments. In Marinos PG, Koukis GC (eds) The engineering geology of ancient works, monuments and historical sites, International Symposium, Athens 10–23 September 1988. A.A. Balkema, Rotterdam/ Brookfield, pp 1063–1069

Chapman S (2003) Building Egyptian pyramid – achieving the impossible (1st Books Library). Authorhouse, Bloomington, IN

Crozat P (2002) Le genie des pyramides. Ed. Dervy, Paris

Cuer M, Bayer R (1980) Fortran routines for linear inverse problems. Geophysics 45:1760

Demortier G (2009) Revisiting the construction of the Egyptian pyramids. Europhysics News 40(1):26–31

Dormion G (2004) La Chambre de Cheops. Fayard, Paris

Dormion G, Goidin JP (1986) Kheops. Nouvelles enquetes. Editions Recherche sur les Civilisations

Gondran M, Vergnieux R (1997) Amenophis IV et les pierres du soleil. Akhenaton retrouve. Arthaud, Paris

Goyon G (1977) Le secret des batisseurs des grandes pyramides. Pygmalion, Paris

Guerrier E (2006) Les pyramides, l'enquete. Cheminements Ed

Hawass Z (2003) La fantastique histoire des batisseurs de pyramides. Traduction par C. Hachet. Editions Rocher

Houdin JP (2006) Khufu, The secrets behind the building of the Great Pyramid. Preface by Z. Hawass, Farid Atiya Press

Houdin JP (2009) La pyramide de Kheops revelee. Abydos Publications, Gizeh

Houdin JP, Houdin H (2002) La Construction de la pyramide de Kheops. Annales des Ponts & Chaussees 101:78–83

Kerisel J (1991) La pyramide a travers les ages. Arts et religions. Presses des Ponts & Chaussees, Paris

Kiner A (2007) La preuve par les cavites. Sciences et Avenir, April 2007.

Lakshmanan J (1990) Traitement et inversion des donnees gravimetriques. La microgravimetrie et son application aux recherches de vides. These de l'Universite de Nancy

Lakshmanan J, Erling JC (1987) La prospection microgravimetrique dans la pyramide de Kheops. Annales de l'Institut Technique du Batiment et des Travaux Publics 454, Serie Architecture et Urbanisme:115–122

Lauer JPh (1988) Le Mystere des pyramides. Presses de la Cite, Paris

Lehner M (1985) The Pyramid Tomb of Hetep-Heres and the Satellite Pyramid of Khufu. Deutsches Archäologisches Institut Abteilung Kairo

Lheureux Ph, Marin S (2008) Le mecanisme secret de la grandre pyramide. [The secret mechanism of the Great pyramid]. Editions du temps present, Muret

Maragioglio V, Rinaldi CA (1965) L'Architettura della Piramide Menefite, Rapallo 1965, Vol. IV, Tavola

Montlucon J, Wadier Y, Lefebvre JP, Lapointe T, Deletie P, Martinet A (1987) Etude geomecanique de la chambre du Roi dans la pyramide de Kheops. Notes du groupe de travail d'EDF

Petrie WMF (1892) Ten year digging in Egypt – 1881 1891. Whitefriars Press, London

Rousseau J (2001) Construire la grande pyramide. L'Harmattan, Paris

Safon C, Vasseur G, Cuer M (1977) Some applications of linear programming to the inverse gravity problem. Geophysics 42:1215

Salençon J (1977) Applications of the theory of plasticity in soil mechanics. Wiley, Chichester, New York, Brisbane, Toronto

Permissions and Acknowledgements

Fig. 1.1: Petrie sequence (W. M. F. Petrie, 1892)
Fig. 1.2: Summital platform (After J. Rousseau, 2001)
Fig. 1.5: Meidoum Pyramid (Permission of J.P. Houdin)
Fig. 1.6: Statuette of Cheops (Cairo Museum)
Fig. 2.1: Chromite deposit (Artist view after Mironov)
Fig. 3.8: Cheops pyramid (Tatiana's photo, http ://tatiana.blogs.com)
Fig. 3.18: Babel Tower of P. Brughel (Permission of KHM Vienna)
Fig. 3.20b: Internal ramp tunnel (Permission of J.P. Houdin)
Fig. 4.1: Mythological papyri (After Piankov and Rambowa)
Fig. 4.2: Babel Tower (After J. Kerisel, 1991)

Index

A
Ahmed Kadry, 21, 73
A posteriori, 20
A priori knowledge, 18, 20, 22, 35, 37–38, 43–44, 71
Arabesque, 63–65
Art, 37
Athens symposium, 4, 42, 44–46
Attenuated radon transform, 3
Aufrere, S., 58

B
Babylon tower, 47, 54, 75
Backus and Gilbert method, 18
Bardot, J., 5, 62
Blind test, 21–23, 42, 73
Boat, solar, 4, 26–27, 38, 42, 73
Borchardt, L., 50, 55–60
Bouchard, 4
Bouguer anomaly, 14, 18, 37–38
Bouguer correction, 14, 28, 36
Butterfly effect, 17–19

C
Cenotaph, 5
Champollion, 4
Chevron, 6, 47, 54
Chromite deposit, 15
Complementarity, 75
Conical radon's transform, 3, 72
Convex analysis, 20, 36, 41
Corbelling, 5, 56, 62
Covariant matrices, 37
Crozat, P., 60
Cube sugar, 32

D
Darmon, F., 5, 62
Deficit of gravity, 17, 30
Degree, 6–9, 29, 31, 38, 46–47, 50, 52–61, 63, 67–68, 71–73, 76
Densitogram, 42, 46–48, 50–51, 58–63, 70, 73, 76
Density field, 18, 20, 26
Detection, 1, 3, 25–26, 72
Djedi, 48, 53–54, 64, 74
Drilling, 4, 14–15, 19–20, 38–39, 42, 73

E
Elevator, 60
Excess of gravity, 33
Expectation, 37
Exploration, 1, 3–4, 15–17, 19, 21, 25–27, 34, 38, 42, 73

F
Fitzclarence, G. A. F., 32, 61
Frustration, 75–76
Functional, 20, 35–37, 72

G
Ghost solution, 18, 39
Golden rule/number, 17, 48, 63–65
Gondran, M., 19, 46, 75
Goyon, G., 10, 33, 52–53, 56
Gravimeter, 14, 71
Gravimeter with cold atom, 71
Graviton, 13, 71
Guerrier, E., 50, 58–59, 61, 73, 76

H
Hawass, Z., 4, 10
Hemiounou, sequence, 4, 10
Herodotus, 10–11, 51–52, 54, 56–57, 60, 62, 75–76
Heterogenization, 58
High precision balance, 14–15

Holscher zigzag ramp, 32
Homogenization, 58, 75–76
Homogenized density, 46
Houdin, J. P., 2, 9, 32, 47, 50, 56, 60–61, 63, 73, 76

I
Ill-posed problem, 18
Imaging, 3, 26, 34–35, 39–40, 42–46, 51–52, 60, 62, 68–69, 72
Inequalities constraints, 36, 71
Instability, 33, 58, 72–73
Interstices, 8, 31–33, 38, 46, 68, 76
Inverse problems, 2–3, 17–20, 22, 28, 34–38, 43–44, 50, 58–59, 61, 72
Inversion of gravity, 26, 35–36

J
Japanese team, 19
JOT, 65

K
Ka, 5
Karmarkar, 18
Kerisel, J., 9–10, 32–33, 47, 51, 53–54, 56, 72–74
Khephren, 8–9, 29
King's chamber structure, 27–29, 41

L
Lacoste and Romberg, 14, 26
Layer, 6–7, 9, 17, 22, 29, 32–33, 52, 54, 56–57, 60–61, 76
Layout, 61
Le Patriote, 19

M
Macrocosm, 58
Macro-elements, 41–44, 58, 68
Macroscopic, 57–58, 76
Mean density, 6, 18, 20, 28–31, 33, 35–37, 40, 42–44, 46, 51, 54, 62, 67–69
Measurement campaign, 26–27
Media failure, 4, 20–21
Medical imaging, 3, 34, 51, 72
Meidoum pyramid, 8–9, 31, 33, 52, 54, 58, 73
Menke method, 18, 20
Meshes, 22, 41–43
Microcosm, 58
Micro-elements, 41–42, 46, 50, 58
Microgal, 26
Microscopic, 57–58, 76
MRI, 3, 34
Mystery, 4–5, 10, 51–52, 62–63

N
Newton's law, 13
Nonlinear system, 36
Notch, 32, 50, 60–61

O
One-to-one mapping, 21–22
Open tomb, 5
Optimization problem, 36

P
Papyrus, 52–54, 57
Papyri, 5, 61
Petrie sequence, 6–10, 54, 57, 72
Pneusol, 10
Puzzle of stones, 7–10, 29
Pythagoras(P.) triplet, 53, 74–75
echography, 19, 26

R
Radiography, 3, 26, 39
Radon's transform, 3, 72
Raising, 48–50
Rally games, 36
Ramp tunnel, 60–61, 73
Reconstruction, 19, 46–47, 51–65
Reinforcement, 10, 38, 72
Residual, 36–37, 40–41, 43
Rousseau, J., 51, 56

S
Sabatier, Paul, 17
Sabatier, Pierre, 37
Salençon, J., 20, 33, 72
Sauneron, S., 10–11
Scanner, 3, 26, 34, 39, 46
Season card, 39
Simplex, 18
Soil mechanics, 33
Solar boat, 4, 26–27, 38, 42, 73
Spiral, 44–47, 53, 60–61, 63–65, 76
Stability, 6–7, 33, 58, 67, 72–73
Stacking, 72
Stage (of construction), 55, 57
Statuette (of Cheops), 10–11
Sunflower, 63–64

T
Talatat, 19, 46
Tarantola, 20, 37, 71

Tatiana, 34, 59
Tikhonov, 18, 37
Titanic, 19
Tomography (structural/functional), 72

U
Uniqueness, 50, 61

V
Vergnieux, R., 19, 46
Voids, 27, 29, 31–33, 40, 46, 59, 62–63, 67, 72, 76

W
Weyl, H., 25
WLS method, 18, 37

MIX
Papier aus verantwortungsvollen Quellen
Paper from responsible sources
FSC® C105338

If you have any concerns about our products,
you can contact us on
ProductSafety@springernature.com

In case Publisher is established outside the EU,
the EU authorized representative is:
Springer Nature Customer Service Center GmbH
Europaplatz 3, 69115 Heidelberg, Germany

Printed by Libri Plureos GmbH
in Hamburg, Germany